CONTINENTS ADRIFT

Readings from
**SCIENTIFIC
AMERICAN**

CONTINENTS ADRIFT

with introductions by
J. Tuzo Wilson
University of Toronto

W. H. Freeman and Company
San Francisco

Most of the SCIENTIFIC AMERICAN articles in
Continents Adrift are available as separate
Offprints. For a complete list of more than 800
articles now available as Offprints, write to
W. H. Freeman and Company, 660 Market Street,
San Francisco, California 94104.

Printed in the United States of America

Library of Congress Catalog Card Number: 73–190438

Standard Book Number: 0–7167–0858–2 (cloth)
0–7167–0857–4 (paper)

9 8 7 6 5 4 3 2 1

This book brings together fourteen articles from SCIENTIFIC AMERICAN that describe the latest scientific revolution—the revolution in ideas about the behavior of the earth's surface. Formerly, most scientists regarded the earth as rigid and the continents as fixed, but now the surface of the earth is seen as slowly deformable and the continents as "rafts" floating on a "sea" of denser rock. The continents have repeatedly collided and joined, repeatedly broken and separated in different patterns, and, very likely, they have grown larger in the process.

This scientific revolution, as others before it, was long in the making, but it was not until the late 1960s that it began to succeed. At a meeting of the world's geophysicists in August of 1971, it was made clear that the notion of continental drift, which had been heresy only a few years before, had become the orthodoxy of the great majority.

The articles in this book, which trace the history of this remarkable change, are of three sorts. Those in the first section were written before their authors had been converted to a belief in continental drift, but they deal with discoveries that, nevertheless, made acceptable the idea that the earth is being slowly deformed. The second group of articles deals with the establishment of the idea of sea-floor spreading and reveals the intellectual struggle attendant upon the abandonment of old beliefs in favor of the new. The third section describes some of the consequences of this scientific revolution.

In calling this reversal of opinion a scientific revolution, I am following those historians who hold that most branches of science developed out of the practical experience of such men as miners, seafarers, farmers, and foundrymen. At first, each branch of science was no more than a codification of the knowledge accumulated over many generations; as such, it offered few surprises and provided little basis for prediction. Historians of science—notably T. S. Kuhn—have pointed out that, as the quantity of knowledge increases, each branch of science reaches a stage in which theoreticians reinterpret the lore of practical men into new and subtler formulations. Many of these have seemed, at first sight, to be contrary to reason, but because they are consistent with the body of accumulated observation, they have been accepted, in time, as superior interpretations.

In his book *The Structure of Scientific Revolutions* (2nd ed.; The University of Chicago Press, 1970), Kuhn cites, as a classical example of a scientific revolution, the discard of the Ptolemaic belief that the earth is the center of the solar system in favor of Copernican astronomy. To think of the earth as the center of the universe and to consider that it requires a fixed support is the obvious interpretation; to realize that the earth is spinning freely in space, and that the sun —not the earth—is the center of man's universe, was a prodigious achievement. This new enlightenment opened the way for Galileo, Kepler, and Newton, who provided science with a burst of great discoveries. Lest we dismiss too lightly the Ptolemaic astrologers, however, we should recall that they devised the calender we use today, invented methods by which eclipses can be predicted accurately, and were the navigators on the voyages of discovery to the New World and on the first circumnavigation of the globe.

Similarly, the change from phlogistic alchemy to modern chemistry, the change from a belief in many separate creations to a belief in organic evolution, and the acceptance of quantum mechanics and relativity in modern physics were scientific revolutions that led to great developments without the abandonment of older observations.

Today, many earth scientists believe that, within the past decade, a scientific revolution has occurred in their own subject. As before, the new beliefs do not invalidate past observations; the new beliefs depend upon reinterpretations of geology and geophysics, and they demonstrate the interdependence of the two disciplines. The acceptance of continental drift has transformed the earth sciences from a group of rather unimaginative studies based upon pedestrian interpretations of natural phenomena into a unified science that is exciting and dynamic and that holds out the promise of great practical advances for the future.

December 1971   J. Tuzo Wilson

CONTENTS

Note on cross-references: References to articles included in this book are noted by the title of the article and the page on which it begins; references to articles that are available as Offprints, but are not included here, are noted by the article's title and Offprint number; references to articles published by SCIENTIFIC AMERICAN, but which are not available as Offprints, are noted by the title of the article and the month and year of its publication.

I

MOBILITY IN THE EARTH

I

MOBILITY IN THE EARTH

INTRODUCTION

The concept of movable continents has been with us for more than a century, but is was not widely accepted until recently: traditionally, earth scientists have regarded the earth as rigid. The articles in this section do not deal directly with the hypothesis of continental drift, but, rather, with discoveries that led to the realization that the earth's interior may be slowly deformable. One possible consequence of a deformable interior, of course, is a mobile surface.

During the four centuries since the shores of the Atlantic were first mapped, many observers—including Francis Bacon, Placet, and Buffon—discussed theories of its formation. Some, including von Humboldt, noted the similarity of the shapes of the opposing coasts of Africa and South America. The comments of these men were vague and brief, however, and it was not until 1958 that a Frenchman, A. Snider, clearly set forth the concept that the continents had once been fitted together in a single supercontinent and had subsequently moved apart. Few could comprehend how continents could possibly move through ocean floors of solid rock, and most ignored Snider's ideas.

Only gradually did geologists begin to move away from the notion of a rigid earth. They observed the great folding and shortening of the strata in alpine mountains and concluded that the two sides of each such range of mountains had necessarily moved together. Geophysicists, by indirect instrumental means, discovered that this compression had created deep roots of light surface rocks that effectively buoy up the mountains. They also observed that Scandinavia, from which great ice sheets have recently melted, is rising, and they reasoned that the maintenance of such vertical motion requires some horizontal movement of material. They concluded that some part of the earth's interior must be mobile, which would allow for horizontal movements of the surface. Between 1910 and 1912, Frederick B. Taylor, H. D. Baker, and Alfred L. Wegener all advanced views about continental drift quite similar to those that are widely held today. The outbreak of World War I delayed discussion of their arguments, but between 1915 and 1929, Wegener published four editions of his book *Die Entstehung der Kontinente und Ozeane*, which were translated into several languages, including English (*The Origin of Continents and Oceans*). Wegener's theories provoked a great controversy. Several geologists in Europe (notably E. Argand, A. Staub, and A. Holmes) and others in the Southern Hemisphere (led by A. L. du Toit and, later, A. Maach, S. Warren Carey, and A. Ahmad) agreed with him, but most North American geologists did not, with the notable exception of W. A. J. M. van Waterschoot van der Gracht. Most geophysicists agreed with Harold Jeffreys that drift was physically impossible, and, for a generation, only a handful of geologists actively supported the idea. One group of geologists working in South America and Africa considered that the strata and fossils of one continent so closely resemble those of the other that they once must have formed parts of one great supercontinent (they called it Gondwanaland) that has since broken apart.

Because Jeffreys continued vigorously to point out that the mechanics of moving solid rock through solid rock appeared to be impossible and that the cause of the supposed motion was unknown, the discoveries discussed in the papers of this first section were essential to the acceptance of the theory of continental drift.

The first article, "The Origin of the Earth," was written by the Nobel laureate Harold C. Urey. He concluded that the old idea that the solar system developed sequentially in a gaseous state—with the planets deriving from the sun, and the moon from the earth—is wrong. He postulated that all parts of the system formed at the same time from a giant dust cloud similar to many still observable in the universe. (Today, versions of the hypothesis that the earth had a cold origin and warmed up later are universally accepted.) During the early history of the earth, gravitational energy and the decay of radioactive elements, which were then more abundant, provided sufficient heat to cause melting. Urey suggested that only the metallic components melted and that they migrated toward the center to form the core; others believe that the whole earth became molten. If so, the mantle has solidified again; it cannot have cooled far below its melting point, however, because once it solidified, heat would only be able to escape slowly. It is possible that this has an important bearing on the mechanism of continental drift, because solids are deformable at temperatures close to their melting points. Urey also discussed certain geochemical problems, including the formation of the oceans. They have certainly played a part in fashioning the continents, for it can hardly be a matter of chance that the ocean basins are just the proper depth to hold the seas and the continents are just high enough to be dry land.

Robert L. Fisher and Roger Revelle, in their article "The Trenches of the Pacific," provided an exciting account of early exploration of the great trenches that ring the Pacific Ocean and are by far its deepest parts. The article describes the earthquakes and the active volcanoes that follow the trenches, and points out that little sediment has accumulated in them and that their vicinity is marked by a deficiency in gravity, which indicates that they are underlain by light material. It also notes that a seamount on one slope of the Tonga Trench has sunk and become tilted. From these indications, the authors concluded that motions acting against the force of gravity must be pulling the crust down beneath the trenches and dragging sediments into the earth. They did not know the nature of those forces, and they did not then believe in continental drift; yet their discoveries support our present belief that trenches form over subduction zones, where one moving plate of the earth's crust overrides another that is sinking into the mantle.

Marshall Kay's article "The Origin of Continents" is next. This article is important because Kay suggested therein that ocean trenches and island arcs evolve into geosynclines and mountain belts, and that the volcanic rocks brought to the surface along such arcs may add to the continents, causing them to grow. Some details of the theory have been modified by later work, and no one yet knows for certain how much of the crust and ocean waters was formed very early in the earth's history and how much has been added since, but it does seem likely that the continents, the sea, and the atmosphere have developed and changed with the passage of geological time.

The last two papers in this section discuss the nature of the earth's interior. The first of these, K. E. Bullen's "The Interior of the Earth," shows that the interior may be divided into a series of concentric shells, the outermost of which (Bullen's "layer A") is the crust. A sharp boundary called the Mohorovičić Discontinuity separates the crust from the mantle. It is universally agreed that the composition of the crust is different from that of the mantle: the crust is granitic (and about 20 miles thick) beneath the continents and basaltic (and about 3 miles thick) beaneath the oceans, whereas the uppermost 250 miles of the mantle (which forms Bullen's "layer B") is probably largely composed of olivine, which is a magnesium iron silicate.

Don L. Anderson, in his article "The Plastic Layer of the Earth's Mantle," which was published seven years later, agreed in general with Bullen, but made the important additional point that there is a plastic layer in the upper mantle upon which the continents may move. Anderson revived the old concepts of lithosphere and asthenosphere and pointed out that these layers are not the same as the crust and mantle. The Mohorovičić Discontinuity is a sharp boundary that marks the change in chemical composition between the more siliceous crust and the more basic mantle. The distinction between the lithosphere and the asthenosphere, by contrast, is a matter of strength, rather than chemistry. The lithosphere is cool enough to be rigid and strong. It extends from the surface to an average depth of about 40 miles; thus, it includes the whole crust and the top of the mantle as well. Although not deformable, the lithosphere is brittle and can break into plates or spherical caps. The hot asthenosphere beneath it is believed to be deformable: if it has no permanent strength, the lithospheric plates can easily slide over it, and their vertical motions can account for isostasy. Most likely, the asthenosphere is at least partially molten. This means that it is a source of much lava, and it explains why earthquake waves travel more slowly through it than through either the lithosphere or the deeper mantle.

1

THE ORIGIN OF THE EARTH

HAROLD C. UREY
October 1952

IT IS PROBABLE that as soon as man acquired a large brain and the mind that goes with it he began to speculate on how far the earth extended, on what held it up, on the nature of the sun and moon and stars, and on the origin of all these things. He embodied his speculations in religious writings, of which the first chapter of *Genesis* is a poetic and beautiful example. For centuries these writings have been part of our culture, so that many of us do not realize that some of the ancient peoples had very definite ideas about the earth and the solar system which are quite acceptable today.

Aristarchus of the Aegean island of Samos first suggested that the earth and the other planets moved about the sun— an idea that was rejected by astronomers until Copernicus proposed it again 2,000 years later. The Greeks knew the shape and the approximate size of the earth, and the cause of eclipses of the sun. After Copernicus the Danish astronomer Tycho Brahe watched the motions of the planet Mars from his observatory on the Baltic island of Hveen; as a result Johannes Kepler was able to show that Mars and the earth and the other planets move in ellipses about the sun. Then the great Isaac Newton proposed his universal law of gravitation and laws of motion, and from these it was possible to derive an exact description of the entire solar system. This occupied the minds of some of the greatest scientists and mathematicians in the centuries that followed.

Unfortunately it is a far more difficult problem to describe the origin of the solar system than the motion of its parts. The materials that we find in the earth and the sun must originally have been in a rather different condition. An understanding of the process by which these materials were assembled requires the knowledge of many new concepts of science such as the molecular theory of gases, thermodynamics, radioactivity and quantum theory. It is not surprising

GREAT FURROW (*right center*) on the surface of the moon must have been made by a tough metallic object. The author believes that this was a fragment of a large body that crashed into the moon from the right.

that little progress was made along these lines until the 20th century.

The Earlier Theories

It is widely assumed by well-informed people that the moon came out of the earth, presumably from what is now the Pacific Ocean. This was proposed about 60 years ago by Sir George Darwin. The notion was considered in detail by F. R. Moulton, who concluded that it was not possible. In 1917 it was again considered by Harold Jeffreys, who thought that his analysis indicated the possibility that the moon had been removed from a completely molten earth by tides. In 1931, however, Jeffreys reviewed the subject and concluded that this could not have happened; since then most astronomers have agreed with him.

But although Moulton and Jeffreys showed the improbability of the origin of the moon from the earth, they proposed theories for the origin of the solar system involving the removal of the earth and the other planets from the sun. Together with James Jeans and T. C. Chamberlin they proposed that another star passed near or collided with the sun, and that the loose material resulting from this cosmic encounter later coagulated into planets. This idea of the origin of the solar system has been widely held right up to the present.

The evidence gathered by our great telescopes now tells us that most of the stars in the heavens are pairs or triplets or quadruplets. We have determined the masses of multiple stars by means of Newton's laws of motion and his universal law of gravitation; we have also studied the velocities of these stars by significant changes in their spectra and by actually measuring the motions of nearby examples. We find that the two stars of a pair seldom have exactly the same mass, and that the ratio of the mass of one star to that of the other varies considerably. Gerard P. Kuiper of the University of Chicago concludes that the number of pairs of stars is entirely independent of the ratios of their masses; that is, there is very little probability that one ratio of masses would occur more often than another. In fact, it would appear that there is about as much chance of finding a pair of stars in which one has one-thousandth the mass of the other as there is of finding a pair in which one is 999 thousandths as massive as the other.

Of course it would be very difficult to see a double star in which the secondary was only a thousandth as large as the primary, particularly if the second emitted no light. The sun and Jupiter, the largest of the planets, might be viewed as such a double star: Jupiter weighs about a thousandth as much as the sun, and it shines only by reflected sunlight. Even from the nearest star Jupiter would

CLOUD OF DUST from which the solar system evolved may have developed this intricate pattern of turbulence, suggested by the German physicist C. F. von Weizsäcker. The dust in each eddy gradually coagulated.

be invisible. There is much evidence, however, that a double star such as the sun and Jupiter should occur as a regular event in our galaxy, and the same considerations would seem to indicate that there may be as many as a hundred million solar systems within it. Solar systems are almost certainly commonplace, and not the special things that one might expect from the collision of two stars.

The Dust Cloud Hypothesis

Many years ago E. E. Barnard of the Yerkes Observatory observed certain black spots in front of the great diffuse nebulae that occur throughout our galaxy. Bart J. Bok of Harvard University has investigated these opaque globules of dust and gas; they have about the mass of the sun and about the dimensions of the space between the sun and the nearest star. Lyman Spitzer, Jr., of Princeton University has shown that if large masses of dust and gas exist in space, they should be pushed together by the light of neighboring stars. Eventually, when the dust particles are sufficiently compressed, gravity should collapse the whole mass, and the pressure and temperature in its interior should be enough to start the thermonuclear reaction of a star.

It would seem reasonable to believe that if a star such as the sun resulted from a process of this kind, there might be enough material left over to make a solar system. And if the process was more complex we might even end up with two stars instead of one. Or again we might have triple stars or quadruple stars. Theories along this line are more plausible to us today than the hypothesis that the planets were in some way removed from the sun after its formation had been completed. In my opinion the older hypotheses were unsatisfactory because they attempted to account for the origin of the planets without accounting for the origin of the sun. When we try to specify how the sun was formed, we immediately find ways in which the material that now comprises the planets may have remained outside of it.

One piece of evidence that must be included in any theory about the origin of the solar system consists in our observation of the angular momentum that resides in the spinning sun and the planets that travel around it. The angular momentum of a planet is equal to its mass times its velocity times its distance from the sun. Jupiter possesses the largest fraction of the angular momentum in the solar system; only about two per cent resides in the sun. Another fact that must

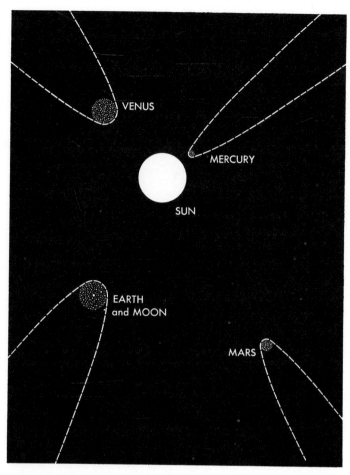

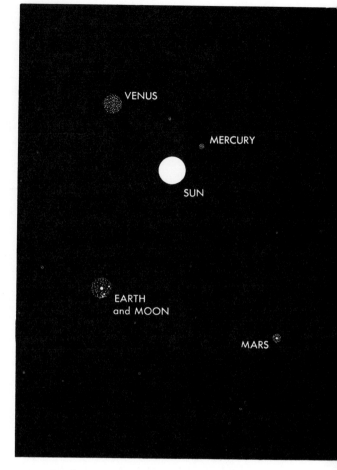

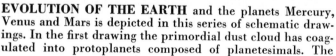

EVOLUTION OF THE EARTH and the planets Mercury, Venus and Mars is depicted in this series of schematic drawings. In the first drawing the primordial dust cloud has coagulated into protoplanets composed of planetesimals. The gases that have coagulated with the planetesimals are driven away (*dotted lines*) by the pressure of light from the sun. In the second drawing the gas has been completely removed from the proto-

be encompassed by any theory is the so-called Titus-Bode law, which points out in a simple mathematical way how the distances of the planets from the sun vary: the inner planets are closer together and the outer ones are farther apart. This is only an approximate law which does not hold very well, and perhaps more emphasis has been put upon it than it deserves. In my own study of the problem I have looked for other evidence regarding the origin of the solar system.

Some 15 years ago Henry Norris Russell of Princeton and Donald H. Menzel of Harvard pointed out that there was a very curious relationship between the proportions of the elements in the atmosphere of the earth and the atmospheres of the stars, including the sun. It is particularly noteworthy that neon, the gas that we use in electric signs, is very rare in the atmosphere of the earth but is comparatively abundant in the stars. Russell and Menzel concluded that neon, which forms no chemical compounds, escaped from the earth during a hot early period in its history, together with all of the water and other volatile materials that constituted its atmosphere

at that time. The present atmosphere and oceans, they proposed, have been produced by the escape of nitrogen, carbon and water from the interior of the earth. The German physicist C. F. von Weizsäcker similarly suggested that the argon of the air has resulted mostly from the decay of radioactive potassium during geologic time, and has escaped from the interior of the earth. F. W. Aston of Cambridge University also pointed out that the other inert gases, krypton and xenon, were virtually missing from the earth.

The Chemical Approach

My own studies in the origin of the earth started with such thoughts about the loss of volatile chemical elements from the earth's surface. Exactly how did these elements escape from the earth, and when? I came to the conclusion that it was impossible that they were evaporated from a completely formed earth; the evaporation must have occurred at some earlier time in the earth's history. Once the earth was formed its gravitational field was much too strong for volatile gases to escape

into space. But if these gases escaped from the earth at an earlier stage, what is the origin of those that we find on the earth today? Water, for example, would have tended to escape with neon, yet now it forms oceans. The answer seems to be that the chemical properties of water are such that it does not enter into volatile combinations at low temperatures. Thus if the earth had been even cooler than it is today, it might have retained some water in its interior that could have emerged later. But meteorites contain graphite and iron carbide, which require high-temperatures for their formation. If the earth and the other planets were cool, how did these chemical combinations come about?

Indeed, what was the process by which the earth and other planets were formed? None of us was there at the time, and any suggestions that I may make can hardly be considered as certainly true. The most that can be done is to outline a possible course of events which does not contradict physical laws and observed facts. For the present we cannot deduce by rigorous mathematical methods the exact history that began with a globule of dust. And if we cannot

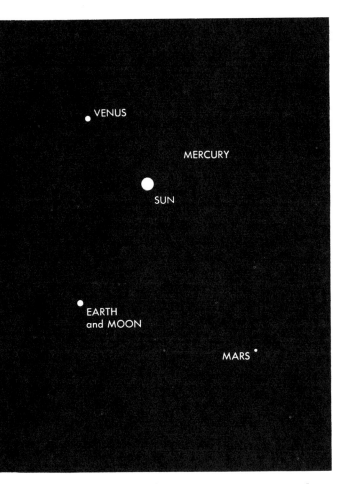

planets. In the third drawing the planetesimals have formed the planets. The relative sizes of the sun and planets and the distances between them have been distorted for purposes of diagrammatic clarity.

do this, we cannot rigorously include or exclude the various steps that have been proposed to account for the evolution of the planets. However, we may be able to show which steps are probable and which improbable.

Kuiper believes that the original mass of dust and gas became differentiated into one portion that formed the sun and others that eventually became the planets. The precursors of the so-called terrestrial planets—Mercury, Venus, the earth and Mars—lost their gases. The giant planets Jupiter and Saturn retained the gases, even most of their exceedingly volatile hydrogen and helium. Uranus and Neptune lost much of their hydrogen, helium, methane and neon, but retained water and ammonia and less volatile materials. All this checks with the present densities of the planets.

It seems reasonably certain that water and ammonia and hydrocarbons such as methane condensed in solid or liquid form in parts of these protoplanets. The dust must have coagulated in vast snowstorms that extended over regions as great as those between the planets of today. After a time substantial objects consisting of water, ammonia, hydro-

carbons and iron or iron oxide were formed. Some of these planetesimals must have been as big as the moon; indeed, the moon may have originated in this way. The accumulation of a body as large as the moon would have generated enough heat to evaporate its volatile substances, but a smaller body would have held them. Most of the smaller bodies doubtless fell into the larger; Deimos and Phobos, the two tiny moons of Mars, may be the survivors of such small bodies.

Massive chunks of iron must also have been formed. On the moon there is a huge plain called Mare Imbrium; it is encircled by mountains gashed by several long grooves. It would seem that the whole formation was created by the fall of a body perhaps 60 miles in diameter; this has been suggested by Robert S. Dietz of the U. S. Naval Electronics Laboratory, and by Ralph B. Baldwin, the author of a book entitled *The Face of the Moon*. The grooves must have been cut by fragments of some very strong material, presumably an alloy of iron and nickel, that were imbedded in this body. Of course large objects of iron still float through interplanetary space; occasionally one of them crashes into the earth as a meteorite.

How were such metallic objects made from the fine material of the primordial dust cloud? In addition to dust the planetesimals contained large amounts of gas, mostly hydrogen. I suggest that the compression of the gases in a contracting planetesimal generated high temperatures that melted silicates, the compounds that today form much of the earth's rocky crust. The same high temperatures, in the presence of hydrogen, reduced iron oxide to iron. The molten iron sank through the silicates and accumulated in large pools.

It now seems that the meteorites were once part of a minor planet that traveled around the sun between the orbits of Mars and Jupiter. The pools of iron that formed in this body may have been a few yards thick. In the case of the object that was responsible for Mare Imbrium and its surrounding grooves, the

depth of the pools must have been several miles. If the temperature of such a planetesimal had been high enough, its silicates would have evaporated, leaving it rich in metallic iron. The object must eventually have cooled off, for otherwise its nickel-iron fragments could scarcely have been hard enough to plow 50-mile grooves on the surface of the moon.

It was at this stage that the planetesimals lost their gases; Kuiper believes that they were probably driven off by the pressure of light from the sun. This left the iron-rich bodies that are today the earth and the other planets. The whole process bequeathed a few meaningful fossils to the modern solar system: the meteorites and the surface of the moon, and perhaps the moons of Mars.

The Moment of Inertia

Recently we have redetermined the density of the various planets and the moon. The densities of some, calculated at low pressures, are as follows: Mercury, 5; Venus, 4.4; the earth, 4.4; Mars, 3.96, and the moon, 3.31. The variation is most plausibly explained by a difference in the iron content of these bodies. And this in turn is most plausibly explained by a difference in the amount of silicate that had evaporated from them. Obviously a planet that had lost much of its silicate would have proportionately more iron than one that had lost less.

It is assumed by practically everyone that the earth was completely molten when it was formed, and that the iron sank to the center of the earth at that time. This idea, like the conception of an earth torn out of the sun, and a moon torn out of the earth, almost has the validity of folklore. Was the earth really liquid in the beginning? N. L. Bowen and other geologists at the Rancho Santa Fe Conference of the National Academy of Sciences in January, 1950, did not think so. They argued that if the earth had been liquid we should expect to find less iron and more silica in its outer parts.

There is other evidence. Mars, which should resemble the earth in some respects, contains about 30 per cent of iron and nickel by weight, and yet we have learned by astronomical means that the chemical composition of Mars is nearly uniform throughout. If this is the case, Mars could never have been molten. The scars on the face of the moon indicate that at the terminal stages of its formation metallic nickel-iron was falling on its surface. The same nickel-iron must have fallen on the earth, but there it would have been vaporized by the energy of its fall into a much larger body. Even so, if the earth had not been molten at the time, some of the nickel-

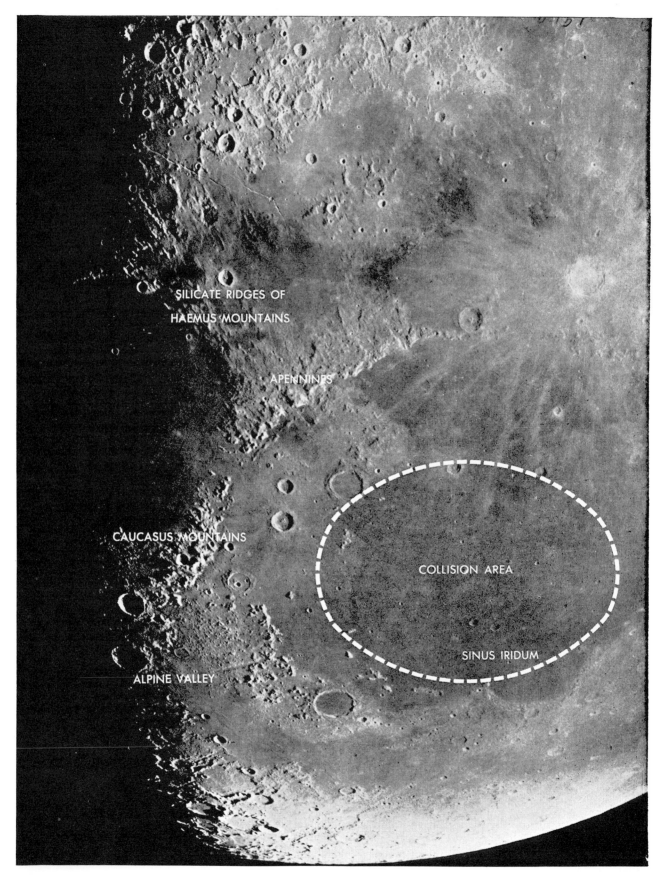

SILICATE RIDGES OF
HAEMUS MOUNTAINS

APENNINES

CAUCASUS MOUNTAINS

COLLISION AREA

SINUS IRIDUM

ALPINE VALLEY

MARE IMBRIUM, a large circular plain near the northern edge of the moon, was probably made by the fall of a body about 60 miles in diameter. This planetesimal apparently ploughed in through Sinus Iridum and spread out in a collision area outlined by relatively small, iceberg-like masses. The rocky silicates of the planetesimal splashed out in the region of the Haemus and the Apennine Mountains. The Alpine Valley, which is the same formation that appears on the preceding page, may have been made by a metallic object.

iron might still be found in its outer mantle.

If there is iron in the mantle of the earth, it may be sifting toward the center of the earth; and if it is moving toward the center of the earth, it will change the moment of inertia of the earth. The moment of inertia may be defined as the sum of the mass at each point in the earth multiplied by the square of the distance to the axis of rotation, and added up for the whole earth. If iron were flowing toward the interior of the earth, this quantity should decrease. It is a requirement of mechanics that if the moment of inertia of a rotating body decreases, its speed of rotation must increase. Finally if the speed of the earth's rotation is increasing, our days should slowly be getting shorter.

Now we know that our unit of time is changing; but it is getting longer, not shorter. That is, the earth is not speeding up but is slowing down. Very precise astronomical measurements, some of them dating back to the observation of eclipses 2,500 years ago, indicate that the day is increasing in length by about one- or two-thousandths of a second per day per century. It has been thought that the lengthening of the day was due to the friction of the tides caused by the sun and the moon. But if we attempt to predict changes in the apparent position of the moon on the basis of this effect alone, we find that our calculations do not agree with the observations at all. If on the other hand we assume that iron is sinking to the core of the earth, the changing moment of inertia would also influence the length of the day. Indeed, calculations made on the basis of both the tides and the changing moment of inertia do agree with the observations.

In order to make the calculations agree we must postulate a flow of 50,000 tons of iron from the mantle to the core of the earth every second. Staggering though this flow may seem, it would take 500 million years to form the metallic core of the earth. Some calculations indicate that it may have taken as long as two billion years. The important thing is that the order of magnitude approaches that of the age of the earth, which is generally given as two to three billion years. If this reasoning is correct, the earth was made initially with some iron in its exterior parts, and it could not have been completely molten.

To complicate matters Walter H. Munk and Roger Revelle of the Scripps Institution of Oceanography have shown that the moment of inertia of the earth is probably decreasing because water is slowly being transferred from the oceans to the ice caps of Greenland and Antarctica, and that this process can account for the lengthening of the day without assuming that iron is moving to the center of the earth, at least not so rapidly as I have calculated. In view of the argument of Munk and Revelle we really have no evidence for the flow of iron to the center of the earth. However, we have little evidence to the contrary. New observations are needed.

The Last Stages

Let us briefly retell what the course of events may have been. A vast cloud of dust and gas in an empty region of our galaxy was compressed by starlight. Later gravitational forces accelerated the accumulation process. In some way which is not yet clear the sun was formed, and produced light and heat much as it does today. Around the sun wheeled a cloud of dust and gas which broke up into turbulent eddies and formed protoplanets, one for each of the planets and probably one for each of the larger asteroids between Mars and Jupiter. At this stage in the process the accumulation of large planetesimals took place through the condensation of water and ammonia. Among these was a rather large planetesimal which made up the main body of the moon; there was also a larger one that eventually formed the earth. The temperature of the planetesimals at first was low, but later rose high enough to melt iron. In the low-temperature stage water accumulated in these objects, and at the high-temperature stage carbon was captured as graphite and iron carbide. Now the gases escaped, and the planetesimals combined by collision.

So, perhaps, the earth was formed!

But what has happened since then? Many things, of course, among them the evolution of the earth's atmosphere. At the time of its completion as a solid body, the earth very likely had an atmosphere of water vapor, nitrogen, methane, some hydrogen and small amounts of other gases. J. H. J. Poole of the University of Dublin has made the fundamental suggestion that the escape of hydrogen from the earth led to its oxidizing atmosphere. The hydrogen of methane (CH_4) and ammonia (NH_3) might slowly have escaped, leaving nitrogen, carbon dioxide, water and free oxygen. I believe this took place, but many other molecules containing hydrogen, carbon, nitrogen and oxygen must have appeared before free oxygen. Finally life evolved, and photosynthesis, that basic process by which plants convert carbon dioxide and water into foodstuffs and oxygen. Then began the development of the oxidizing atmosphere as we know it today. And the physical and chemical evolution of the earth and its atmosphere is continuing even now.

2

THE TRENCHES OF THE PACIFIC

ROBERT L. FISHER AND ROGER REVELLE
November 1955

On April 28, 1789, Lieutenant William Bligh, commanding H.M.S. *Bounty*, had a memorable quarrel in the Pacific Ocean with his senior warrant officer, one Fletcher Christian, as a result of which they parted company and sailed off in opposite directions—Christian in the *Bounty* and Bligh in the ship's longboat. This historic mutiny took place near the great volcano of Tofua in the Friendly Islands, better known today as the Tonga Islands. Bligh and Christian were well acquainted with the fact that the oceanic topography around these islands was somewhat unusual—full of treacherous shoals and narrow interisland passages. But they could not know, for methods of deep-sea sounding had not yet been invented, how unusual it really was, nor that this place would one day yield one of the most remarkable discoveries in the history of ocean-going exploration.

Beneath the placid sea east of the Tonga Islands yawns a monstrous chasm nearly seven miles deep. A hundred years after the *Bounty* episode another British vessel first plumbed its depths. Surveying the ocean bottom around the islands, Pelham Aldrich, commanding H.M.S. *Egeria*, was surprised to find that on two occasions his sounding lead did not touch bottom until 24,000 feet of wire had been paid out. Aldrich's discovery prompted other nations to send out expeditions to explore the Tonga undersea abyss. Eventually they traced out a great trench running from the Tonga Islands south to the Kermadec Islands [*see map on page 14*]. The deepest sounding made recently by the research vessel *Horizon* of the Scripps Institution of Oceanography, is some 35,000 feet. The immense chasm plunges about 6,000 feet farther below sea level than Mount Everest rises above it!

The Tonga-Kermadec Trench is now known to be but one member in a vast chain of deep, narrow trenches which lie like moats around the central basin of the Pacific [*see map*]. All of them parallel island archipelagoes or mountain ranges on the coasts of continents. Along the coast of South America, the drop from the top of the Andes to the bottom of the offshore trench is more than 40,000 feet. And the length of the undersea troughs is no less remarkable than their depth: some are 2,000 miles long.

These great gashes in the sea floor are so unlike anything on land that they are difficult for us as land animals to visualize. It is hard to grasp the reality of a chasm so deep that seven Grand Canyons could be piled on one another in it, and so long that it would extend from New York to Kansas City. Yet these are the dimensions of the Tonga-Kermadec Trench.

The size and peculiar shape of the Pacific trenches stir our sense of wonder. What implacable forces could have caused such large-scale distortions of the sea floor? Why are they so narrow, so long and so deep? What has become of the displaced material? Are they young or old, and what is the significance of the fact that they lie along the Pacific "ring of fire"—the zone of active volcanoes and violent earthquakes that encircles the vast ocean?

Although the trenches are still only sketchily explored, some tentative answers to these questions can be gleaned from the information already obtained. We can take the Tonga-Kermadec Trench as a typical example.

The Trench lies on a long, nearly straight, north-south line east of the Tonga and Kermadec archipelagoes. At its northern end it has a slight hook. It begins there as a gentle, spoon-shaped depression, runs southeasterly between Tonga and Samoa, then turns and deepens, strikes south for 1,200 miles and finally shoals and disappears at a point

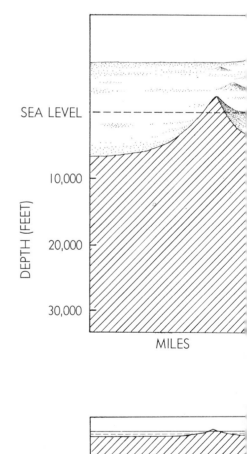

TONGA TRENCH would appear as in the upper drawing if viewed northward from a point in the central Tonga Islands, and if vertical distances were exaggerated by a

north of New Zealand. In its deepest central portion the Trench is very narrow—no more than five miles wide. The chasm is V-shaped, but the arm of the V on the island side is considerably steeper than on the seaward side: on the landward western wall the slopes average from 16 to 30 per cent—*i.e.*, in places they are steeper than the 24-per-cent-average slope of the sides of the Grand Canyon at Bright Angel. In longitudinal section the Trench consists of deep depressions separated by saddles; it looks like beads on a string, or peaks and saddles on an upside-down mountain range.

The islands on the western lip of the Trench appear to be part of the same crustal structure. They lie in two lines on a thousand-mile-long ridge atop the Trench's western slope. The islands of the Polynesian kingdom of Tonga are capped with limestones, laid down in shallow water during the last era of geologic time. These islands rest on broad shelves of drowned coral, 180 to 360 feet deep, and they rise in a series of terraces to a few hundred feet above sea level. West of the limestone islands, separated from them by a shallow trough, is a chain of submarine volcanoes and high volcanic islands. The volcanoes are explosive, rather than of the quiet Hawaiian variety. They have contributed great quantities of ash and cinders to the surrounding sea floor. Five of the island volcanoes have erupted during the last hundred years, and the danger of further explosions has forced the government of Tonga to evacuate the inhabitants. There are also active volcanoes below the sea surface. One of them, Falcon Bank, rises several hundred feet above the sea during an eruption; indeed, this bank is commonly called Falcon Island. After each eruption waves quickly erode the erupted lava, and within a few years the volcano is submerged again.

The floor of the Tonga-Kermadec Trench is rocky and seems to be nearly bare of sediments. During the Scripps Institution *Capricorn* expedition of 1952-1953, a core barrel with a heavy lead weight, which because of difficulties with the winch was dragged along the sea floor for several hours before it could be raised, came up badly battered by the bottom rocks. The heavy steel bail holding the instrument had been bent, and the lead weight looked as if it had been beaten with a hammer and chisel. Small fragments of black volcanic rock were embedded in the lead.

On the seaward slope of the Trench a single volcanic cone rises smoothly 27,000 feet, to within 1,200 feet of the sea surface. Just below its summit is a broad flat bench, tilted to the westward. Further study of this great cone, one of the highest mountains on earth, might tell us much about the history of the trench. Almost certainly the flat bench was cut by waves when the topmost part of the peak was above sea level. If

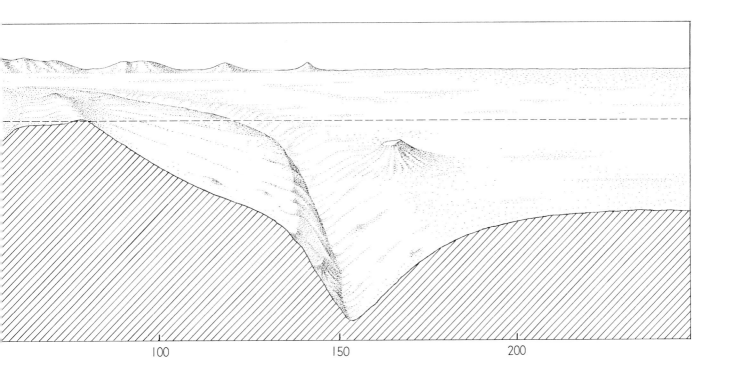

factor of 10 in comparison with horizontal. (In the lower drawing the cross section is shown without vertical exaggeration.) The exposed land mass in the left foreground is the island of Kao, a dormant volcano. In the distance are the Samoan Islands. The flat-topped seamount on the eastern slope of the trench is one of the highest mountains in the world. Its summit, which is tilted down toward the west about one degree, was probably worn flat by wave action when trench was shallower and mountain above water.

shallow-water fossils could be recovered from the summit, we could fix the time when submergence occurred, and perhaps when the bench began to be tilted. This in turn might give us information about the rate of downward bending of the trench floor.

The Tonga Trench, as we have said, is typical of the great trenches in the Pacific. Some of the other giant furrows are the Aleutian, Kurile, Japan, Marianas, Philippine and Java Trenches on the northern and western sides and the Acapulco and Peru-Chile Trenches on the eastern side of the ocean. It is a remarkable and probably significant fact that the deepest trenches all have about the same maximum depth. The record sounding so far was one estimated to be somewhere between 35,290 and 35,640 feet, made in the trench southeast of the Mariana Islands. Appropriately enough this depth was measured by H.M.S. *Challenger*, the modern namesake of the famous ship whose voyage around the world in the 1870s marks the beginning of modern oceanography [see "The Voyage of the *Challenger*," by Herbert S. Bailey, Jr.; SCIENTIFIC AMERICAN Offprint 830]. The original *Challenger* actually discovered the Marianas depression, and for many years it was known as the Challenger Deep.

All the deep trenches seem to be generally V-shaped in cross section, although some are slightly flattened at the very bottom; in the Japan and Philippine trenches this flat portion is two to 10 miles wide. Some shallower trenches and trenchlike depressions are U-shaped, with extremely flat bottoms over broad areas, as if they had been partly filled with sediments. If the V-shaped trenches contain any sediments, the layer cannot be more than a few hundred feet thick.

Direct exploration of the trenches is most difficult. Their great depth and extreme narrowness present formidable obstacles. To lower a dredge or other heavy sampling apparatus to the bottom of the deeper trenches, the ship needs a tapered wire rope of the strongest steel and a powerful, specially designed winch. Only three such winches exist today. One was built for the Swedish *Albatross* Expedition of 1948-49 and was later used on the Danish *Galathea* Expedition of 1950-52; another is installed on the Scripps Institution's research vessel *Spencer F. Baird*; the third is on the U.S.S.R. research ship *Vitiaz*. The winch drum on the *Baird* carries 40,000 feet of wire rope. When this wire was paid out in the Tonga Trench with

a heavy core barrel on the end, the strain at deck level was 12 tons.

A single lowering of a dredge or core barrel takes many hours. It is complicated by the problem of keeping a small ship in position in a rolling sea, often under the influence of strong and unpredictable currents and shifting winds. The hazards of fouling the wire or of machinery breakdown under the heavy strain are always present, with the possible loss of the precious cable. Such a loss would be crippling, and much of the investment in time and effort required to send a scientific ship to a remote part of the world would also be lost.

If sounding and sampling of the bottom are difficult, drilling to find what lies beneath the bottom of the trenches is quite impossible, with present techniques. For such explorations we must depend on indirect methods—studies of earthquake waves, measurements of gravity anomalies, the flow of heat through the crust and the magnetic properties of the buried rocks.

The zone of trenches is the scene of our planet's most intense earthquake activity. Nearly all the major earthquakes, especially those originating at great depths, occur in this zone. The deepest-focus earthquakes are associated with the deepest and steepest trenches. This strongly suggests that the seat of the trench-producing forces lies far below the earth's surface.

The earthquakes may, indeed, be responsible for the fact that a line of explosive volcanoes parallels the trenches. Some investigators have proposed that the heat produced near the focus of an earthquake melts the surrounding rocks, and that the melted material rises and is eventually ejected by the volcanoes.

Seismic refraction studies give us another clue to the nature of the crust under the trenches. These investigations have shown that beneath the trenches (Tonga and others) the outer crust is less than one third as thick as under the continents. We therefore arrive at the important conclusion that the crustal structure under the trenches is oceanic and not continental.

The most striking phenomenon associated with the trenches is a deficiency in gravity. The force of gravity depends on the mass of matter between the surface and some great depth in the earth. In general this force at any given latitude is about the same in ocean basins as in the continents, despite the fact that the volume of rock under a continental area is greater than under an equal area

of the ocean. Evidently the continents "float" high above the deep sea floor, like rafts of light material in a heavier medium. Within the continents themselves, there is usually little difference in gravity between high mountains and low plains, and it is commonly supposed that the mountains are underlain by a larger thickness of light material than the plains. This state of the earth's crust is called isostatic equilibrium.

Measurements of gravity near trenches show pronounced departures from the expected values. These gravity anomalies are among the largest found on earth. It is clear that isostatic equilibrium does not exist near the trenches. The trench-producing forces must be acting against the force of gravity to pull the crust under the trenches downward!

What may these forces be? Here studies of heat flow in the crust suggest a possible answer. It has long been known that there is a small, steady flow of heat from the earth's depths outward toward the surface. Most of this heat is generated by the disintegration of radioactive elements in the crust and the mantle just beneath the crust. Near the surface of the earth the heat is transported outward principally by conduction, but at greater depths there may be a slow upward movement of the hot rock itself, carrying heat toward the surface. If rock at these depths moves upward in some regions of the earth, there must be other regions where cold rock moves downward. This movement would reduce the outward flow of heat. Now measurements near the floor of the Acapulco Trench show that the flow of heat there is less than half the average for the earth's surface (the average being about 250 calories per year per square inch of surface). So it may be that relatively cool rocks are slowly moving downward under the trench. Such a downward flow would tend to drag the crust down with it and may well account for the formation of the trench. If this process is occurring, the earth's mantle should be cooler under the trench than elsewhere. Magnetic measurements suggest that this is in fact the case, but they are too few so far to be conclusive.

Speculating from what we know, we may imagine that a trench has the following life history. Forces deep within the earth cause a foundering of the sea floor, forming a V-shaped trench. The depth stabilizes at about 35,000 feet, but crustal material, including sediments, may continue to be dragged downward into the earth. This is suggested by the

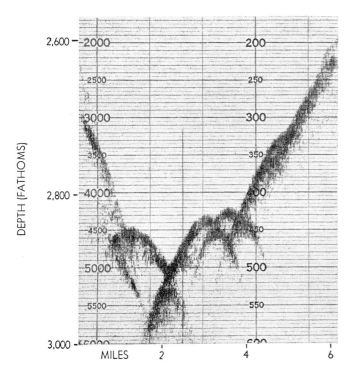

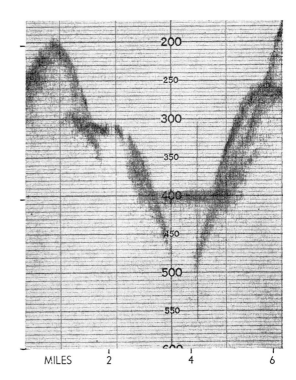

ACAPULCO TRENCH is revealed in cross section by echo sound-ings. Near Acapulco (*left*), the bottom is V-shaped, with little sedi- ment, and 2,930 fathoms deep. Near Manzanillo (*right*), it is flat at 2,795 fathoms. Numbers on the records do not represent fathoms.

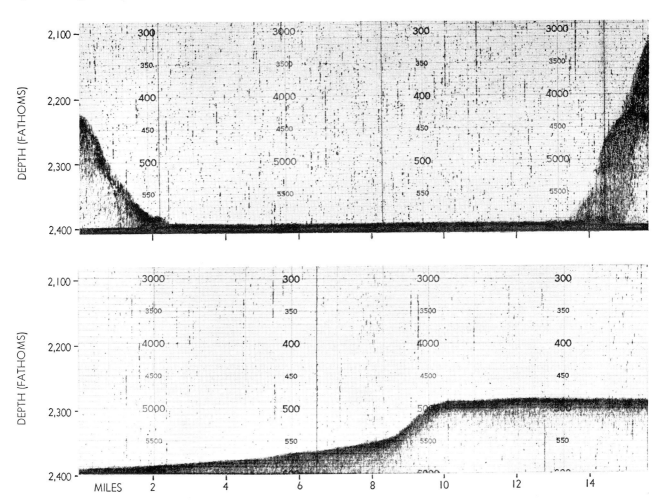

CEDROS TROUGH, a short trench off the coast of Lower Cali-fornia, is traced in cross section by the upper echo-sounding and longitudinally by the lower. The bottom is flat at 2,395 fathoms and measures 11 miles across at the point where measurement was made.

fact that the deepest trenches contain virtually no sediments, although they are natural sediment traps. During this stage in the trench's history there is violent volcanic and earthquake activity.

In a later stage the internal forces pulling or squeezing the crust downward under the trench become less active, and the trench begins to fill up with sediments. It acquires a flat bottom and a U shape as the accumulating sediments cover the topographic irregularities. The sediments may eventually pile up so high that the top of the pile rises above the sea, forming islands, when isostatic equilibrium is finally restored. The topmost sediments will be rock of the kind that is deposited in shallow water, like the limestones of Tonga and the Marianas.

Another process also may come into play if a thick layer of sediments accumulates. Because of their lower heat conductivity, the sediments would form an insulating blanket along the former trench. This would block the heat flow from the interior and cause a temperature rise which would partly melt the deep rocks. The melted material might then move upward and transform the heavy existing rock and the lower part of the sedimentary layer into light, granite-like rock. The thickness of the crust therefore should increase.

Some geophysicists have suggested that it was by such a sequence of events, occurring repeatedly during the geologic past, that the continents grew, at the expense of the ocean basins. The question then arises: Where on the continents are the ancient, filled-in trenches?

One naturally thinks at once of the long, narrow structures, called geosynclines, where sediments piled up and mountain ranges developed by compression and folding. Were some geosynclines originally deep trenches such as now exist on the sea floor? It has usually been thought that this is not so, because most of the sediments in geosynclines appear to have been laid down in shallow water rather than in deep trenches. However, this appearance may in some cases be an illusion. Sediment samples collected from even the deepest trenches resemble in many ways deposits laid down in shallow water.

It is true that the sedimentary rocks in geosynclines contain no recognizable fossils of deep-sea animals. But trenches have little life that could leave a distinctive record. The depths of a trench are completely dark, except for the fee-

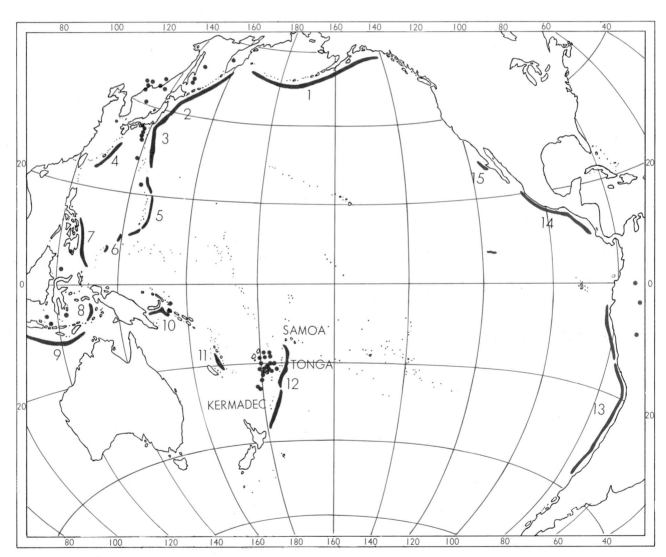

RING OF TRENCHES around the central basin of the Pacific is shown in color. The colored dots represent deep centers of earthquake activity. Numbered serially are: Aleutian Trench (1), Kurile Trench (2), Japan Trench (3), Nansei Shoto Trench (4), Marianas Trench (5), Palau Trench (6), Philippine Trench (7), Weber Trough (8), Java Trough (9), New Britain Trench (10), New Hebrides Trench (11), Tonga-Kermadec Trench (12), Peru-Chile Trench (13), Acapulco-Guatemala Trench (14), Cedros Trough (15).

ble and flickering light produced by luminous organisms, and no plants can live there. The animals and bacteria of the abyss must gain their sparse food supply from plant and animal remains settling slowly from the upper layers of the sea. The waters are very cold: about 36.5 degrees Fahrenheit now, though they may have been some 20 degrees warmer in the geologic past. The pressure at the bottom of a trench is, of course, enormous—more than eight tons per square inch.

The Danish *Galathea* Expedition several years ago dredged up a few animals from the floors of trenches more than 30,000 feet deep. The principal animals recovered were sea cucumbers and a type of sea anemone, neither of which would be likely to leave distinctive fossils. Some worms, clams and crustaceans also were obtained, together with beautiful glass sponges.

Materials which are usually supposed to be deposited only in shallow water have actually been found on the floor of some of the deep trenches. The *Galathea* recovered fine gray sand, pebbles, cobbles and land-plant debris from the floor of the Philippine Trench. The Lamont Geological Observatory of Columbia University found, in cores from the Puerto Rico Trough, the skeletons of plants and animals that live only at shallow depths. In the flat-bottomed northern part of the Acapulco Trench one core contained soft black mud, high in organic debris and stinking of hydrogen sulfide, while in other cores layers of gray, green and brownish sand and silt were interbedded with charred woody fragments and fine green mud.

However, it is clear that some geosynclines, notably those along what are now the Appalachian Mountains, could not have been deep sea trenches, for they contain deposits from marshes and flood plains interbedded with marine sediments, and therefore the deposits must have been laid down in shallow water.

The question remains: Where are the trenches of yesteryear? Are we living in an exceptional geologic era; are the apparently young trenches of the present day unusual formations that have had no counterparts during most of geologic time? Such a speculation would be repugnant to many geologists, because it would be difficult to reconcile with the doctrine that the present is the key to the past. We must continue to search for ancient trenches—on the deep-sea floor, in the marginal shallow water areas and on the continents themselves.

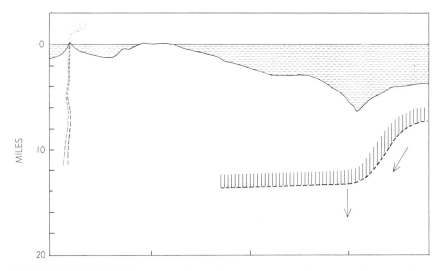

POSSIBLE LIFE HISTORY of a trench is outlined in the three diagrams on this page. Here a force deep within the earth pulls down the floor of the ocean to form a V-shaped trench.

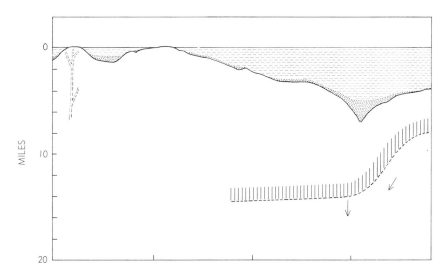

SEDIMENT COLLECTS in the bottom of the trench during the second stage, when the deep force weakens and relaxes its downward pull. Bottom is now flat and V changes to U.

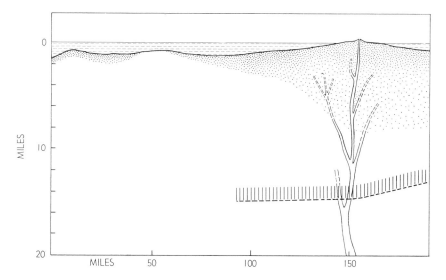

SEDIMENT BUILDS UP and, with isostatic adjustment of the crust, rises above the surface of the sea. Deep molten rock rises through the sediment and is released by volcanoes.

3

THE ORIGIN OF CONTINENTS

MARSHALL KAY
September 1955

The continents and the ocean basins are distinctly different aspects of the earth's crust. Not only do the continents stand higher, but they are made of different material. The difference can be summed up roughly in two chemical terms: sialic and simatic. The continents are composed chiefly of sialic (for silica-alumina) rock, which is especially rich in silica and comparatively light in weight and color; the ocean basins are mainly simatic (silica-magnesium) rock, which is denser and darker. Because granite is the chief sialic rock and basalt the chief simatic one, the continents are most commonly described as granitic and the ocean basins as basaltic.

For decades it has been realized that any attempt to explain the origin of the continents and the oceans must account for this fundamental difference in the respective portions of the earth crust. Many hypotheses have been erected on theories as to how the earth itself was formed. Those who supposed that the earth grew by the coming together of small, cold bodies (planetesimals) reasoned that the difference between the oceanic and continental parts of the crust was due to chemical evolution: the areas covered with water developed differently from those exposed to the atmosphere. The much larger school who believe that the earth was originally a molten mass have suggested a great variety of other explanations. Many have argued that the cooling earth was once crusted over its whole surface with a thin layer of granitic material, and that this layer was later parted in places to expose the basalt underneath—either by gradual drifts of the granitic material which piled it in continents, or by catastrophic events, such as a great oscillation that tore a chunk out of the earth and threw it into

space as the moon, leaving the hole that is now the Pacific Ocean. Others have proposed that the entire earth was originally basaltic, and that the separation of ingredients to form the differentiated oceanic and continental areas was started by deformations of the earth's outer layers. Still others have suggested that the granitic rocks crystallized and floated to the top as the molten earth cooled.

Many of these hypotheses have collapsed in the face of new information about the crust and the interior of the earth obtained during the past decade. The question as to how the continents and oceans were formed is approached by geologists from the opposite direction: instead of starting with speculations about the beginning of the earth and projecting hypotheses from those speculations, they start from the known facts about the present earth and work backward to reconstruct its unknown history. What we know about the continents and oceans gives us a number of clues for deducing their past.

Our firsthand information about the inside of the earth of course is still scanty: it is limited to studies of surface rocks, to soundings and samples of the ocean bottoms and to borings some four miles down into the earth's crust in mines and wells. But in recent years explorers of the interior have probed the earth deeply and intensively with revealing instruments. Foremost among them is the seismograph. Tracing the travel of waves from earthquakes and artificial explosions through the earth, timing their speed and plotting their paths, it has been possible to obtain a kind of X-ray picture of the earth's layered body. And this picture has been confirmed and filled in to some extent by gravity measure-

ments which define areas and belts of differing density.

About three fourths of our planet is covered by ocean waters, but not all of that is actually oceanic basin, for the continents have broad shelves extending far out under a wash of shallow sea. The deep ocean basins—two miles or more below water level—account for about half of the earth's solid surface. These are the areas of the earth's crust that show a sharp contrast to the continents.

The conventional division between the crust and the "interior" of the earth is a boundary known as the Mohorovicic discontinuity—a transition zone between crystalline, basaltic rock and denser, noncrystalline rock beneath. This boundary is not level: it is lower under the continents than under the oceans. In other words, the continents not only stand higher but also plunge deeper into the dense sublayer. It is as if the crust of the earth consists of relatively light but thick continental blocks and thinner but denser oceanic blocks, both of which float on the substratum of the interior. As

SURFACE ROCKS of North America are classified on this map. The distribution of various types provides evidence for hypotheses on origin and growth of the continent.

SEDIMENTARY ROCKS

CRYSTALLINE ROCKS

PROMINENTLY VOLCANIC COVERING OTHER ROCKS

CRYSTALLINE AND SEDIMENTARY ROCKS

VOLCANIC AND SEDIMENTARY ROCKS

measured by the seismic yardstick, the continental blocks are some 30 miles thick and the oceanic basins only about six or eight miles thick. The continental blocks are not granitic all the way down: at their base they have a layer of basalt, like the rock of the ocean basins.

Now to this general picture there are certain exceptions which look extremely significant. Some areas of the earth crust are neither strictly continental nor strictly oceanic—they seem to combine a little of both! These areas are the island archipelagoes: the islands of Japan; the chain consisting of the Philippines, the East Indies and New Guinea; nearer home, the West Indies. The principal islands in such chains have a dominantly continental character: their rock is chiefly granitic and the crust goes deeper than in the ocean basins. Yet each chain also includes parts which are dominantly oceanic; that is, they are underlain by a thin layer of basaltic rock (covered with sediments) and the crust is comparatively shallow. Gravity readings show long, sinuous bands of gravity deficiency, usually along the submarine troughs associated with the island chains. Seismic studies indicate that the island rock

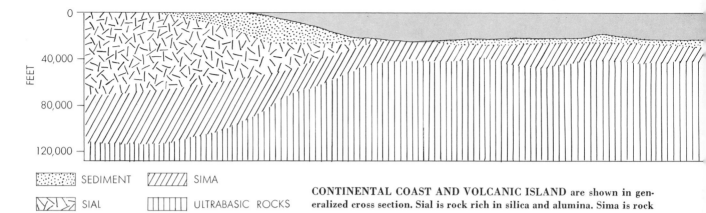

SEDIMENT   SIMA

SIAL   ULTRABASIC ROCKS

CONTINENTAL COAST AND VOLCANIC ISLAND are shown in generalized cross section. Sial is rock rich in silica and alumina. Sima is rock

does not pass beneath the troughs. Apparently the gravity deficiency is due to the thinness of the crust under the trenches rather than to any downfold of granitic rock.

The island archipelagoes, then, are something intermediate between ocean basins and continents. Are they, perhaps, embryo continents in the process of growing up to become larger continents or additions to continents in the future? The question sends us back to examine the rocks of continents with a fresh eye and a hypothesis to investigate. Let us consider the North American continent. What light does the distribution of its various rocks throw on how the continent may have been formed?

The surface of a continent has two general classes of rocks (if we disregard the volcanic rocks, formed from lavas erupted onto the surface here and there). The first great class is the sedimentary rocks—the hardened remnants of sediments laid down by ancient rivers and seas. These are the sandstones, limestones, shales, marls and clays that cover much of our continent. In the central plains they lie in flat layers under the surface soil; in the mountain regions the layers may be tilted, warped, folded and broken, but the rock still has the same character and is unmistakably identified as sediment by the fossil remains of sea animals. These are the rocks that the prospector's drill probes in search of oil and gas. Along the Atlantic Coast the sedimentary rock is only a few thousand feet thick, but at the Gulf Coast in Texas and Louisiana it is piled to a depth of as much as eight to 10 miles.

If we peeled off the sedimentary layers, we would find below, mantling the whole continent, the second great class of rocks. This class includes two types: (1) the "plutonic" or "igneous" rocks, notably granite, and (2) the "metamorphic" rocks—schists, gneisses, quartz-

ites, slates. The metamorphic rocks evidently are sedimentary rocks and lavas altered by heat and pressure. Just how the granites and other igneous rocks are formed is still a matter of debate; there is strong new evidence that they crystallize from molten magma deep in the crust [see "The Origin of Granite," by O. Frank Tuttle; SCIENTIFIC AMERICAN Offprint 819]. In any case, the metamorphic and igneous rocks are generally found together. And wherever we may go on the continents, if we drill deep enough we will find these rocks forming a "basement" beneath the sedimentary cover.

There are three large areas of North America where the granites and metamorphic rocks are not buried but form the surface of the continent. The first is a broad, circular region around Hudson Bay in Canada, extending east to the coast of Labrador and south into Minnesota, Wisconsin and New York where the granitic and metamorphic basement disappears under layers of sedimentary rock and slopes down gradually to its greatest depth at the Gulf Coast. The second belt lies along the Atlantic Coast from central Newfoundland to the Piedmont. The third is on the Pacific Coast from southern Alaska to the tip of Lower California.

This picture is somewhat simplified: parts of the Pacific Coast are covered by sedimentary rocks, and platforms of granite and metamorphic rocks appear in other, smaller, areas of the continent besides the three large ones mentioned. But if we concentrate our attention on these three great areas, we discover some significant relations which bear on the origin of continents.

In the belts along the Atlantic and Pacific Coasts, we can read in these rocks the record of a long series of mountain-building events, marked by the sinking of deep troughs in the crust and the rise of lands nearby. The troughs were deep

enough to catch and pile up miles of sediments and volcanic rock, poured in from adjoining volcanoes. The sediments contain pebbles which must have been washed into them by streams eroding rather rugged lands. And they were invaded by granitic rocks and fluids which transformed them into metamorphic rocks. Moreover, some of the rocks along both the Pacific and Atlantic Coasts have intrusions of very basic rocks (unusually poor in silica) which must have come from considerable depth—peridotites and serpentines, including such rocks as the verd antique marble found in Vermont.

Several lines of evidence indicate that these belts of metamorphosed sediments and volcanic rocks with intrusions are the descendants of belts like the present island chains. (Indeed, the Pacific Coast belt still has an island arc—the Aleutians—extending from its northern end!) So we may deduce that the North American continent has grown on its western and eastern sides by the addition of what were once separate island archipelagoes. The period of upheaval and deformation that formed these additions to the continent seems to have ended, though there is still some unrest along the Pacific Coast. Dating events by the fossils and the radioactivity of rocks, it is judged that the time of principal activity along the Atlantic Coast ended about 200 million years ago, and along the Pacific Coast, perhaps 100 million years ago.

Let us return to the great shield of igneous and metamorphic rocks that covers central Canada. It, too, has granite and granitelike rocks which invaded and altered sediments and lavas. Hence, by analogy, it likewise should once have been an island belt margining the continent. But the rocks are now in the midst of the continent—in fact, as we have seen, the whole continent basement is made of such rocks. How, then, did the continent begin? Did it start as

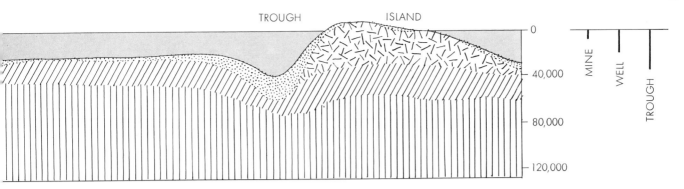

TROUGH   ISLAND

0

40,000

80,000

120,000

MINE   WELL   TROUGH

rich in silica and magnesium. The boundary between sima and ultra-basic rock is called the Mohorovicic discontinuity. At the far right are the depths of the deepest mine (10,000 feet), the deepest well (about 22,000 feet) and the deepest oceanic trough (35,700 feet).

a single small island and grow by addition of island belts? We might be able to answer that question if we had dates for all the intrusive rocks of the continent, so that we could tell whether the central ones are oldest and surrounded by successively younger belts. There do seem to be belts of progressively younger intrusions around the central part of the Canadian shield. But on the other hand there are dated rocks that seem pretty old in far-flung parts of the continent—such as Colorado and Texas. So probably it is not as simple as one might like it to be. A working hypothesis is that the continent has grown from volcanic island belts which were formerly separated by subsiding, sediment-collecting troughs.

It seems that in the beginning all was ocean, from which the continents began to rise as small lands. Perhaps stresses upon the surface of the original whole-ocean earth caused it to buckle, producing troughs, ridges, folds and fissures. The more volatile or acidic or silicic constituents of the fluid rock beneath the surface may have concentrated in the original uplifts. Weathering by the oxygen-rich atmosphere may then have segregated the more acid material of these ridges in sediments. The silicic sediments, deposited along the edges of the

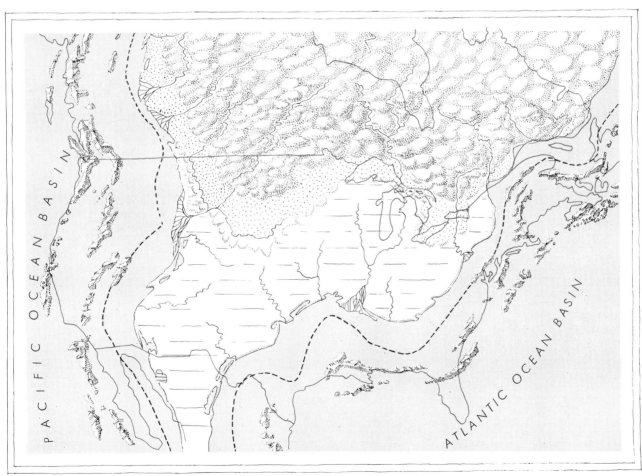

CRYSTALLINE ROCKS (PRE-CAMBRIAN)
CAMBRIAN SANDSTONES
ORDOVICIAN LIMESTONES
INNER LIMIT OF VOLCANIC ROCKS

GEOGRAPHY OF NORTH AMERICA early in the Ordovician Period (about 400 million years ago) is reconstructed on the evidence of today's rocks. Off the east and west coasts of the continent were chains of volcanoes. These were later joined to the land.

raised lands, would have formed the first sialic areas on the earth.

The big question—how the continents became sialic—is full of difficulties and uncertainties. Seemingly the great long-term troughs in the crust of the earth, where thick deposits of sediments and volcanic rocks collect, are in some manner invaded by the underlying magma. Either the rising masses of fluid magma displace these rocks and themselves crystallize as granite, or their fluids and gases transform the existing deposits into granitelike rock. The popular hypothesis for the past decade has been that the furrows plunge so deeply into the substratum that the temperatures and pressures are sufficient to fuse the rocks

and force fluids into them from below. Yet an invasion of this kind can hardly account directly for the production of sialic rock, since the substratum seems extremely dense and low in silica. Perhaps the process is roundabout: first sedimentary rocks, made more silicic than the original rocks by weathering, are laid in the troughs; then they are deeply buried, become fluid and eventually emerge as igneous magma which we no longer recognize as the original sedimentary and volcanic rock.

Any convincing hypothesis of the origin of the continents and oceans must be consistent with a reasonable theory concerning the origin of the earth itself.

Geologists can be judges only of the end of this beginning—the time when the earth began to inscribe a permanent history in rock. The theories of the origin of the earth have been principally the inventions of astrophysicists, celestial mechanics and physical chemists, for the problems lie in those fields. The geologist is concerned only that the result be an earth which conforms to what can be deduced from the earliest known rocks.

Our knowledge of the earth has expanded tremendously in the past decade. Some old hypotheses have had to be abandoned; intriguing new ones are taking their place and will be tested in the next decade by geologists, geochemists and geophysicists all over the world.

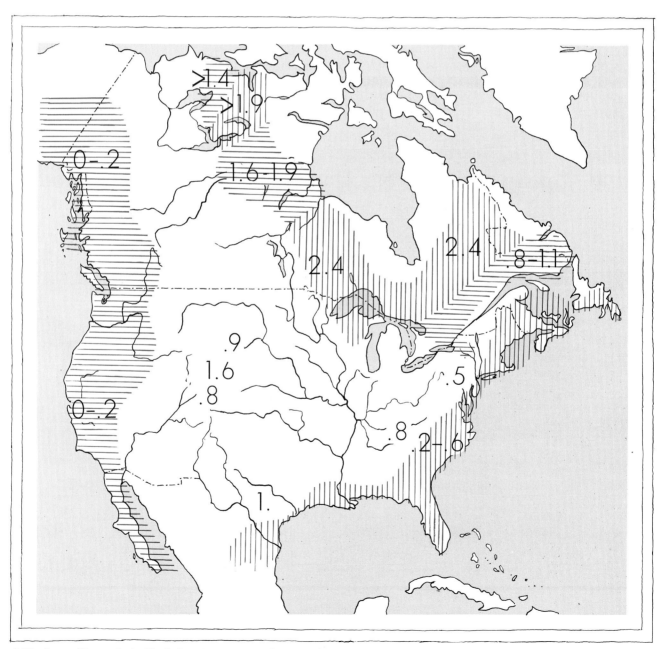

AGE of crystalline rocks in North America is given in billions of years. The oldest rocks are located around Hudson Bay. The young-er rocks are along the present coasts. The symbols before the num-bers at the top indicate that the rocks are older than these values.

4

THE INTERIOR OF THE EARTH

K. E. BULLEN
September 1955

Each year 10 or more major earthquakes shake the earth. The smallest of them releases about a thousand times more energy than an atomic bomb; the Assam earthquake of August, 1950, had about 100,000 times that energy. The waves set up by these convulsions travel through the whole interior of the earth, including the core, and their paths are bent and shaped by the shells of the earth's internal structure. Thus the seismic waves bear clues of the regions they traverse, and from the story they tell when they are received at our seismological stations on the surface it is possible to infer a picture of the interior. In effect the seismologist X-rays the earth, even if at times he sees through a glass, darkly.

Seismology has lifted our notions about the interior of our planet from the realm of wild speculation to the stage of scientific measurement and well-reasoned inferences. Combined with geological information about surface rocks, laboratory experiments on rocks at high pressures and certain astronomical observations, it gives us a basis for learning something about the various conditions in the deep interior—its layered structure, the materials, their physical state, the pressures and so on.

The study of earthquakes is a fairly new science. In 1750 a writer on the subject in the *Philosophical Transactions* of the Royal Society of London apologized to "those who are apt to be offended at any attempts to give a natural account of earthquakes." But observations of earthquake effects accumulated, and late in the 19th century seismology began to emerge as a real quantitative science when the Englishman John Milne constructed in Japan a seismograph suitable for world-wide use. The seismograph was later developed further,

notably by E. Wiechert in Germany, by Prince Galitzin in Russia and recently by Hugo Benioff of the California Institute of Technology.

The release of elastic strain energy at the source, or "focus," of an earthquake produces waves which begin to

radiate in all directions from the focus. In 1897 R. D. Oldham of England identified on seismograms three main types of seismic waves: (1) primary (P) waves, which are compression-and-expansion waves like those of sound; (2) secondary (S) waves, which vibrate at right angles to the direction of travel, as

TWO SEISMOMETERS are photographed at the Lamont Geological Observatory of Columbia University. They consist essentially of three pendulums: one for each dimen-

light waves do; (3) surface waves, which appear in the upper 20 miles or so near the earth's surface. The P waves travel through both solid and liquid parts of the earth; the S waves only through solid.

S waves travel at about two thirds of the speed of P waves. The speed of both varies with depth in the earth; for example, the P waves travel at 8½ miles per second, their maximum speed, at a depth of 1,800 miles and at about three miles per second in rocks near the earth's surface. Because of the change of speed, the path of the waves' travel usually curves upward. When they arrive at a boundary between layers they may be refracted or reflected, and on reaching the earth crust they are reflected downward again. At a boundary either a P or an S wave may give rise to both P and S waves. Thus any one seismogram from a particular earthquake may show many distinct phases, signifying the stages of the waves' routes and their changes of form. A typical seismogram illustrating several phases is shown on the next two pages.

With this kind of evidence Oldham proved in 1906 that the earth has a large central core, and in 1914 Beno Gutenberg, then in Germany, located the boundary of the core at 1,800 miles below the earth's surface. Since the radius of the whole earth is about 3,960 miles, the central core has a radius of some 2,160 miles.

The discovery of the core came about from the observation of shadow zones where relatively few P waves were recorded. Consider P waves issuing from a major earthquake with its focus at the South Pole. These waves would be observed at the surface throughout the Southern Hemisphere and up to 15 degrees above the equator (*i.e.*, the latitude of Guatemala) in the Northern Hemisphere. But between the latitudes of Guatemala and Winnipeg little indication of P waves would be received. Then, from a latitude of 52 degrees north to the North Pole, the waves would come in again strongly. The whole of the U. S. would thus be part of a "shadow zone" for that earthquake. On examination, it was seen that the existence of such

shadow zones required the presence of a central core which would bend sharply downwards the seismic rays striking it from above, somewhat after the manner in which light rays from a stick in water are bent by the water surface.

One of the great labors of seismologists during the first 40 years of this century was to evolve reliable tables for the times of travel of P and S waves along the various phases of their routes. In 1930 Sir Harold Jeffreys of the University of Cambridge, suspecting that the existing "travel-time tables" contained large errors, began a long series of studies to correct them. The author of this article was associated with Jeffreys in this work from 1931 to 1939.

The Jeffreys-Bullen tables of 1940 are now used internationally. They agree closely, in the main, with travel times derived about the same time by Gutenberg and Charles F. Richter at the California Institute of Technology. The travel-time tables are of cardinal importance for charting the structure of the earth's

sion. When the earth shakes, the pendulums tend to stand still. Their apparent motion is then recorded by a pen or a beam of light. One of the pendulums, suspended by a horizontal arm and two diagonal wires, is visible at right in the photograph at left.

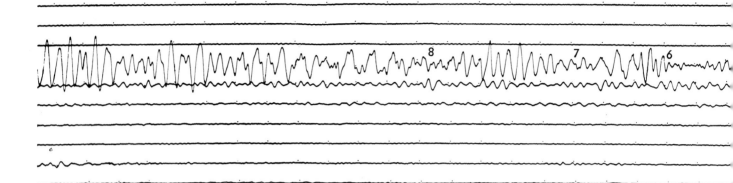

SEISMOGRAM of an earthquake on the Kamchatka Peninsula in Siberia was recorded at the Lamont Geological Observatory. The separate lines are actually part of a continuous spiral trace, going from right to left, on a circular drum. The interval between succes-

interior. It is possible to deduce from the tables the velocities of P waves and S waves in various parts of the interior. Studying the variations of velocity with depth, one can chart different layers and locate boundaries.

With the new tables Jeffreys calculated that Gutenberg's measurement, placing the boundary of the central core at 1,800 miles below the earth's surface, was correct within three or four miles. At least the outer part of the core is judged to be molten. S waves do not pass through it, and its fluid character is established by other evidence, including

data on the tidal deformation of the solid earth and astronomical data on the movements of the earth's poles. H. Takeuchi of Japan has calculated that this region is at most one 300th as rigid as the next outward layer.

The use of the terms "solid" and "fluid" in connection with the huge pressures prevailing in the earth's interior is sometimes questioned. What a geophysicist means by the term "solid" in this context is simply that the elastic behavior of the material in question can be described by equations which match those applying to ordinary solids in

normal laboratory conditions. These equations involve the use of two coefficients: "incompressibility," which is the measure of resistance to pressure, and "rigidity," signifying resistance to shearing stress. In the case of a fluid, the resistance to shear is much smaller than the resistance to compression. This is why a fluid does not transmit S waves.

All the earth outside the core is called the mantle. The whole of the mantle (apart from the oceans and pockets of magma in volcanic regions) is now known to be essentially solid: both S and P waves travel through every part of it.

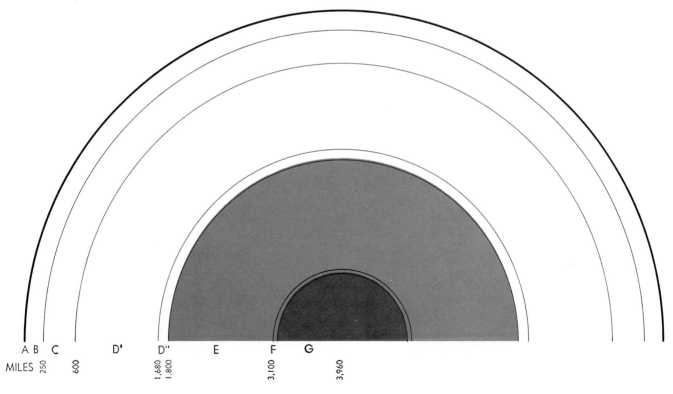

EARTH'S CROSS SECTION is divided into distinct layers through which seismic waves travel at different speeds. The outer core is indicated by the lighter tone of color; the inner core, by the darker tone of color. Layer A is the thin crust of the earth.

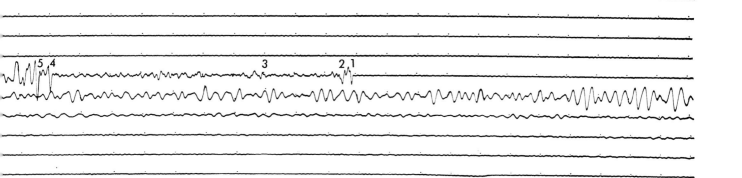

sive dots is one minute. The first disturbance recorded was the P wave designated by the number 1. Then followed the multiply re-flected P waves 2 and 3. S waves begin at position 4, followed by multiply reflected waves at 5, 6 and 7. Surface waves start at 8.

In 1909 the Croatian seismologist A. Mohorovicic, studying the seismograph of a Balkan earthquake, discovered an important discontinuity (boundary) now known to be some 20 miles below the earth's surface. The part of the earth above the Mohorovicic discontinuity has come to be called the crust. But nowadays the term "crust" has only a conventional meaning. According to seismic evidence the crust is not more rigid than the material just below it.

Seismologically speaking, the crust differs from the underlying part of the mantle in the fact that P and S waves travel in it more slowly and with more variable speed. This irregularity of travel velocity makes detailed charting of the crust difficult. The work is being pursued vigorously, however, by the study of surface waves, of P and S waves from near earthquakes (near the recording station), of waves from large man-made explosions (such as the one on Helgoland in 1947) and by seismic probings with dynamite, as in oil prospecting. One important discovery has been that the crust is much thinner under the oceans than under the continents.

Seven distinct regions or shells have now been identified in the earth. In 1936 Miss I. Lehmann of Denmark discovered that the core was not uniform but seemed to consist of at least two different parts. Looking closely at the relatively minor P waves that emerge in the shadow zone on the surface, she concluded that these waves might come to the surface because they were bent sharply upward by an inner core where P waves traveled faster than in the outer core. Her proposal later received support from work by Gutenberg, Richter and Jeffreys. The inner core has a radius of some 800 miles, so the thickness of the

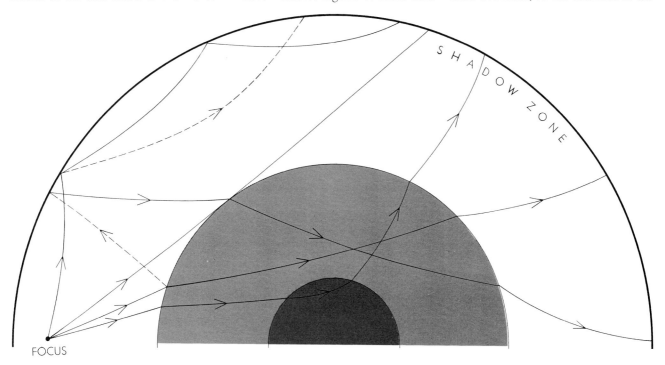

EARTHQUAKE WAVES are bent and reflected as they travel from their source. Solid lines represent P waves, dotted lines are S waves formed by reflection. The only P waves that can get into the shadow zone are those which enter the inner core and are sharply bent.

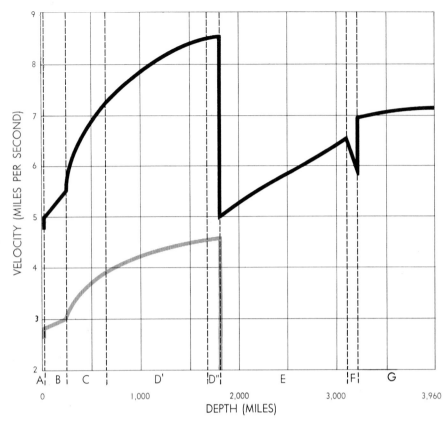

SEISMIC WAVE SPEEDS vary with depth. The black line gives velocities of P waves; the gray line, of S waves. Both change abruptly at core, or E layer, and the S wave disappears.

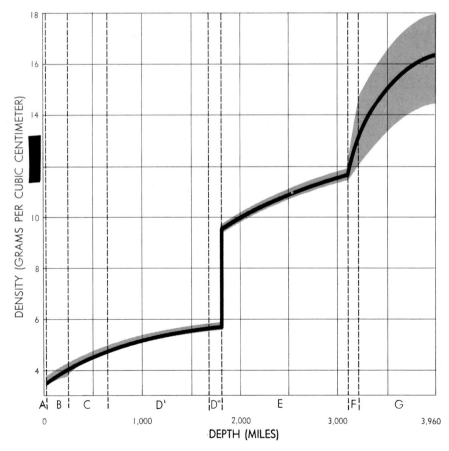

DENSITY of the earth's material increases with depth. The solid line shows the most probable value at each depth and the gray region outlines the probable range of uncertainty.

outer core would be about 1,300 miles.

On the basis of density variations the writer has divided the body of the earth into seven regions, called A, B, C, D, E, F and G [see diagram on page 24]. The A region is the crust. The rest of the mantle below is divided into B, C and D, with D subdivided into D′ and D″. These divisions are still tentative because of certain uncertainties in estimates of velocity gradients. The outer part of the core is called E, and the inner part, G. Between the inner and outer core Jeffreys finds a layer F, some 80 miles thick, where the velocity of P waves declines sharply. Gutenberg has not found this layer, but has said that his data do not exclude its existence.

How can we estimate the pressures and physical characteristics of matter at these various depths in the body of the earth? The velocities of P and S waves are determined by the density, compressibility and rigidity of the material through which they pass, but they do not provide enough information to solve exact equations for those values. There are, however, indirect clues which help us to arrive at estimates—information on the earth's mass and moment of inertia, field observations and laboratory experiments on rocks, mathematical theories of elasticity and gravitational attraction.

By such means the writer has estimated that the earth's density increases gradually from 3.3 grams per cubic centimeter just below the crust to 5½ grams per c.c. at the bottom of the mantle, then jumps suddenly to 9½ grams at the top of the core and thereafter increases steadily to 11½ grams at the bottom of the outer core.

A related calculation gave the increase in pressure with depth in the earth. At the bottom of the Pacific Ocean the pressure is about 800 atmospheres. Only 200 miles down in the mantle the pressure is already 100,000 atmospheres—as great as the highest pressure Percy W. Bridgman of Harvard University has been able to produce in the laboratory. At the base of the mantle, 1,800 miles down, the pressure reaches the immense figure of 1⅓ million atmospheres, and at the center of the earth it is nearly four million atmospheres.

Next the calculations yielded the surprising finding that the rigidity of the material in the mantle increases with depth until, at the mantle's base, it is nearly four times that of steel in ordinary conditions. Below this, in the outer core, the seismic evidence shows that the rigidity sinks to practically zero, meaning that the material is essentially fluid.

Perhaps the most important fruits of this series of calculations have been the findings on compressibility. In spite of the sharp changes in density and in rigidity at the boundary between the mantle and the core, the compressibility of the material does not change substantially at the boundary, according to the calculations. This finding led the writer to examine the theoretical effect of pressures of a million atmospheres or more on materials likely to be present in the core. Taking into account a variety of evidence, the conclusion was that bounds could be set to the compressibilities of materials in the core.

Following this line of argument, it seems highly probable that the inner core, unlike the outer core, is solid in the sense defined. The idea that the inner core is solid, suggested by the writer in 1946 and since developed, would explain the speeding up of P waves when they penetrate into the inner core. Calculation indicates that the inner core is probably at least twice as rigid as steel at ordinary pressures.

On the same line of evidence we can also estimate, as we could not before, the density of the inner core. Apparently at the center of the earth the density is between 14½ and 18 grams per cubic centimeter. Yet another inference is that the increase of density with depth in the inner core (and at the base of the mantle) is greater than average, implying some variability in the composition of that region.

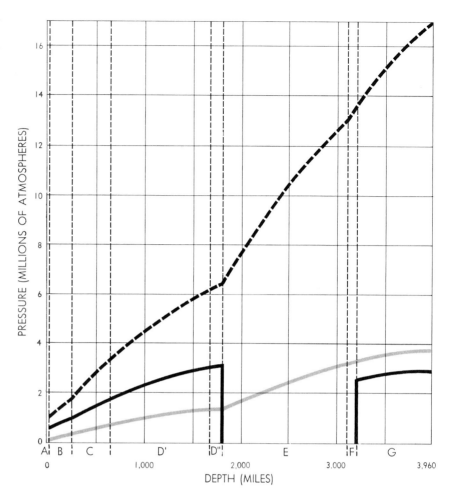

ELASTICITY of the earth's interior is shown graphically above. The gray line measures pressure. Incompressibility is indicated by dotted black line; rigidity, by solid black line.

What is the deep interior of the earth made of? For many years there have been good grounds for believing that much of the mantle below the crust consists of ultrabasic rock such as magnesium-iron silicate. The region B seems to be composed of a material like the known mineral olivine. C appears to be a transition region where the composition changes, perhaps only from one geometric form of olivine to another. The region D' may contain several distinct phases, such as silica, magnesia and iron oxide. The bottom of the mantle, D", is probably of variable composition, but there is as yet no widely accepted agreement on what materials would have gravitated to this depth.

The composition of the central core has lately become the subject of extremely interesting new conjectures. It had long been assumed that the core consists largely of iron or nickel-iron, and this view was supported by analysis of meteorites, believed to be pieces of an exploded planet resembling the earth.

But in 1941 W. Kuhn and A. Rittmann of Germany put forward the radical idea that compressed hydrogen made up the core. This theory, while contradicted by weighty arguments, gave rise to new investigations based on the idea that under increasingly high pressures the material at the base of the earth's mantle might suddenly jump in density. Thus the outer core may consist not of uncombined iron or nickel but of a high-density modification of the rocky material in the mantle just above it. This theory is highly controversial. On balance of probability the present evidence appears to favor a compromise: namely, that the outer core contains both uncombined iron and some material of appreciably smaller atomic number.

An interesting aspect of the new theory is that it makes plausible the idea that the planets Mars, Venus, Mercury and Earth are all of the same primitive over-all composition. Jeffreys has shown that the earth cannot have the same elemental composition as the other planets if its core is completely different from the mantle in composition. According to

calculations by W. H. Ramsey of England and the writer, the observed masses and diameters of Mars and Venus, and the oblateness of Mars, would be accounted for fairly well by the theory that they are composed of terrestrial materials modified by pressure at depth.

As regards the earth's inner core, it probably consists of nickel and iron with perhaps some denser materials as well.

Estimates of the temperatures in the interior of the earth are much less certain than estimates of pressure. In deep mines the temperature rises at the rate of about 30 degrees Centigrade per mile as one descends. If it rose at this rate all the way down to the core the temperature in the center of the earth would exceed 100,000 degrees. Actually it is practically certain that the rate of increase is very much less in the depths of the earth. Present estimates are that the temperature at the center is no more than 2,000 to 6,500 degrees. In any case, it is fairly clear that the increase of temperature in the earth's interior is dwarfed by the increase in pressure.

5

THE PLASTIC LAYER OF THE EARTH'S MANTLE

DON L. ANDERSON
July 1962

Earth scientists have often pointed out that physical conditions inside our own planet are less well understood than those in stars light-years away. Even more paradoxical is the fact that the region within a few hundred miles of the surface presents more problems and gives rise to more technical controversy than the region below. One long-standing item of debate is the zone called the low-velocity layer.

In 1926 the seismologist Beno Gutenberg suggested that earthquake waves slow down when they travel through a zone roughly 100 to 200 kilometers (60 to 120 miles) below the surface. He attributed the effect to a decrease in the rigidity of the material in the zone compared with that above and below it. Most authorities considered his evidence to be dubious at best, and for 30 years they largely ignored his proposal. Recently a mass of data has accumulated that strongly supports the concept of a low-velocity, low-rigidity layer. Its existence has important implications for all theories concerned with structural changes in and near the earth's surface.

The idea that the earth becomes plastic—if not, indeed, liquid—at moderate depths goes back to the earliest days of geology. Volcanic lava flows pointed to a molten interior not too far below the surface. Observations on the rate of increase of temperature in deep mines indicated that if the temperature continues to increase at the same rate, rocks should be molten at depths of less than 100 kilometers. The enormous cracks and folds found in the earth's crust suggested upheavals in a mobile substratum. All this agreed with prevailing views of the origin of the solar system, which held that the earth and other planets had been torn loose from the sun and had had time to solidify only at the surface.

One of the most compelling arguments for some degree of fluidity in the interior came from the principle of isostatic equilibrium. As long ago as 1854 gravity measurements led geologists to suspect that the earth's crust floats on a denser material. Like other floating bodies, the crust seeks an equilibrium, riding deeper where it is heavier and rising higher where it is lighter. Subsequent studies, of both the strength of gravity and the propagation of earthquake waves, confirmed the notion, indicating that mountains have deep roots that support them just as the submerged portion of an iceberg supports the part above water, whereas plains resemble ice floes, having smooth upper and lower surfaces. Moreover, when the load on a part of the crust changes suddenly (on the geological time scale), the surface can be observed to respond by rising or sinking to restore equilibrium. For example, land covered by ice during the last glaciation is still rising at the rate of about a meter per century. Obviously this behavior implies that the material under the crust can flow, if only slowly.

On the other hand, several facts appeared to rule out the idea of widespread fluid material anywhere near the surface. From the tidal distortions of the solid earth in response to the pulls of the sun and moon, Lord Kelvin calculated that the earth is more rigid than steel. Studies of earthquake waves indicated that at depths down to thousands of kilometers the earth transmits not only compression waves (P waves) but also transverse, or shear, waves (S waves). Shear waves, which oscillate at right angles to their direction of motion, cannot propagate through liquids because liquids have no shear strength. When liquids are subjected to shearing forces, they simply flow. Finally, seismologists discovered that earthquakes originate as deep as 700 kilometers below the surface. Since an earthquake represents the abrupt yielding of rock to accumulated stress, it characterizes brittle, not plastic, material.

The answer to this apparent contradiction is suggested by the properties of noncrystalline materials such as glass and pitch, which behave like solids in

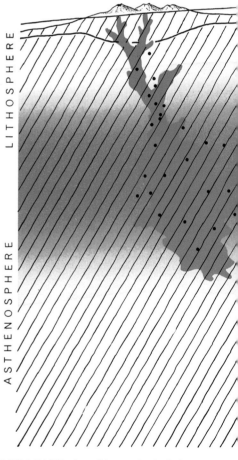

PLASTIC ZONE of earth's mantle (*color*) occupies an ill-defined region some 60 to perhaps 250 kilometers below the surface. In the plastic or low-velocity zone the tem-

the short run and like fluids over longer periods. They transmit shear waves and can support loads for a short time, but under a steady, long-lasting force they are plastic; that is, they flow and change their shape permanently. Under conditions of high temperature and high pressure the rock under the crust could also behave plastically. It would respond like a rigid solid to the relatively short-lived stresses that build up to cause earthquakes and the even briefer stresses involved in earthquake waves, while flowing slowly to adjust to the long-term stresses caused by changes in the weight of overlying material. Some geologists believe that the plastic substance under the crust is a glassy basalt. Recent evidence suggests, however, that it is crystalline. At high temperature even a crystalline material can flow easily, because melting at the boundaries of individual crystal grains allows them to slide over one another.

In 1909 the Yugoslav seismologist Andrija Mohorovičić proposed that at some distance below the surface there is a discontinuity where the velocity of earthquake waves jumps from about seven kilometers per second to eight. Subsequent measurements placed the Mohorovičić discontinuity, or Moho, at an average depth of 35 kilometers below the surface of the continents and only about five kilometers below the ocean floor. Under high mountains the Moho is as deep as 65 kilometers. Geologists saw in the Moho the lower boundary of the rigid, floating crust. The material between the Moho and the presumably liquid core of the earth they named the mantle. Yet the fact that seismic waves travel faster below the Moho than they do above it implies a greater rigidity at the top of the mantle than in the crust. It now seems clear that the Moho marks a change in chemical composition or crystal structure rather than an abrupt transition from strong to weak material.

The first seismic evidence for this transition was not forthcoming until Gutenberg announced the low-velocity zone. Actually what he had found was a decrease in the amplitude of compressional waves reaching the surface at a distance between 100 and 1,000 kilometers from an earthquake. At 1,000 kilometers the amplitudes were only a hundredth as great as they were at 100 kilometers. Beyond 1,000 kilometers the amplitudes increased sharply.

Gutenberg explained the effect by assuming a subsurface layer in which the earthquake waves travel slower than they do in the regions above or below. A wave entering this layer obliquely from above would be refracted downward, away from the surface, as light is bent downward when it passes from air to water. On leaving the bottom of the layer the wave would be refracted upward again [see illustration on page 31]. The result is that the wave would arrive at the surface farther away from its source than it would if there had been no decrease in velocity. Hence a gap would appear between the last "ray" that had missed the low-velocity layer and the first one to enter it. As the illustration shows, the gap, or shadow zone, is great-

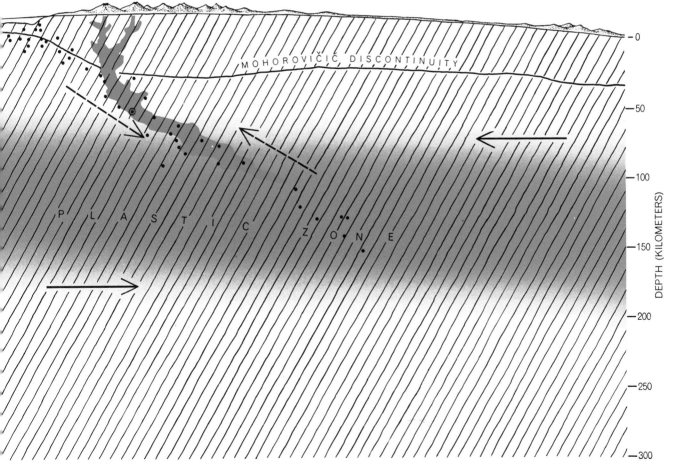

perature approaches the melting point of the rock. The lithosphere is very elastic or brittle rock; the asthenosphere, extending down to earth's core, can flow and relieve stress. In this highly schematic cross section of upper portion of earth, ocean and islands are at left. Dots denote earthquake foci. Broken arrows indicate possible movement of continents over ocean basin. Solid arrows mark hypothetical slippage of whole lithosphere over asthenosphere. The two solid-color regions represent magma, or molten rock.

est for an earthquake originating just above the top of the layer Those coming from deeper levels evince no gap. From the extent of the shadow zone for different earthquakes, Gutenberg calculated that the layer is centered at a depth of about 150 kilometers, and that between 100 and 200 kilometers the velocity is some 6 per cent less than it is just under the Moho. Such a decrease in velocity means that the rock within the layer must be substantially less rigid than the material above and below it. The velocity does not reach the value it had at the base of the crust until some 250 or 300 kilometers below the surface.

If the low-velocity layer were perfectly uniform, and if the waves really traveled as rays, the shadow zone at the surface would be completely "black." No waves at all would emerge within its limits. Actually the layer is full of inhomogeneities, and seismic waves do not travel strictly along classical ray paths. Like all waves, they bend around corners by diffraction, thereby leaking into shadowed regions. Both effects contribute to the energy that is found in the shadow zone.

It was partly this energy leak that made other workers reluctant to accept Gutenberg's conclusion. In those days seismologists paid little attention to the comparative amplitudes of earthquake waves. They were primarily interested in travel times, and they tended to accept any signal, weak or strong, if it appeared in their records at a time when readings at other seismographic stations led them to expect it.

Moreover, the evidence for the low-velocity layer was by no means clear-cut. The statistics were assembled from many earthquakes, large and small, shallow and deep. The data came from seismographs of different designs. In his calculations Gutenberg could make only approximate corrections for these variations as well as for the local irregularities, mostly unmapped, in the rock through which different waves traveled.

Underground nuclear explosions finally made possible a controlled experimental test of Gutenberg's analysis. The time, strength and location of these events is known so precisely that a single blast provides excellent data. Furthermore, seismographs today are more numerous, more sensitive and more standardized than they were in 1926. Studies of several explosions have confirmed the conclusions Gutenberg extracted so tediously from earthquake records [see illustration on page 33]. Seen in sharper detail, the low-velocity layer extends from about 60 kilometers to about 250 kilometers. (It is interesting to note that the layer damps blast waves so effectively that many seismologists think it poses a major difficulty for the detection of underground nuclear tests.)

Several independent pieces of evidence now support the idea of a low-velocity plastic layer. One is furnished by surface waves. These are seismic disturbances that follow the curved surface of the earth [see bottom illustration on page 32] instead of passing through its body. Although the waves travel along the surface, they "feel" the elastic properties of the underlying material to a

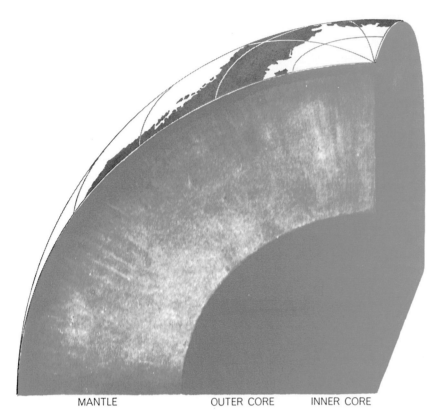

MANTLE   OUTER CORE   INNER CORE

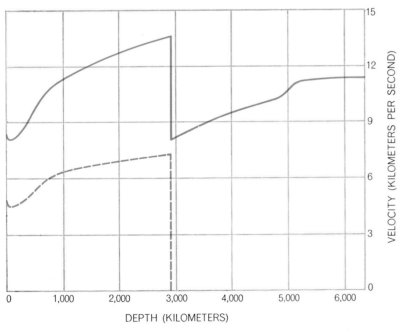

INTERNAL STRUCTURE OF THE EARTH is deduced from travel times of seismic waves. Solid line represents compressional, or P, waves; broken line represents shear, or S, waves. The latter disappear entirely at the outer core, indicating that this region is liquid. Low-velocity zone causes dip in curves at far left. Hatching on block diagram near surface marks low-velocity zone. It also marks transition zone above inner core at depth of 5,000 kilometers.

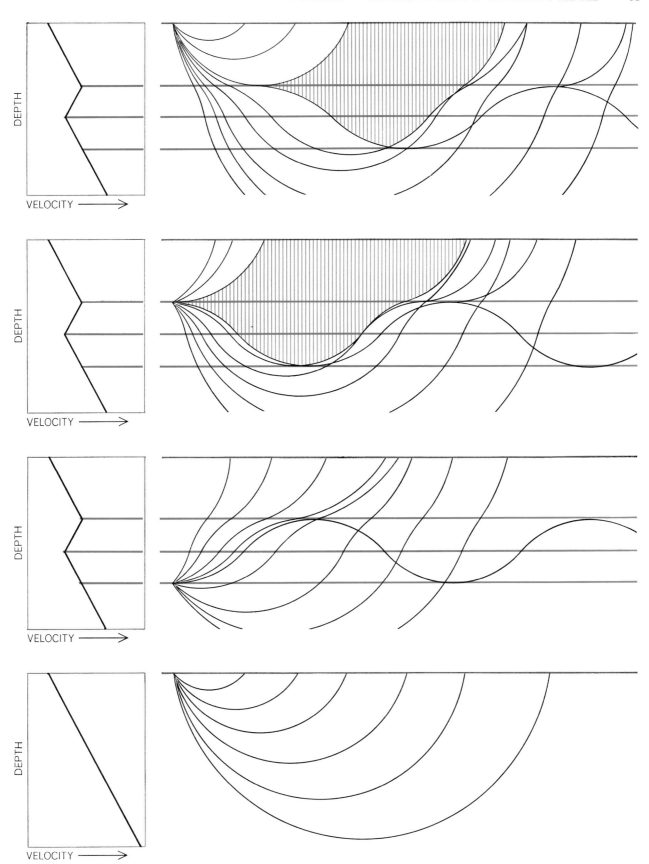

PATHS OF EARTHQUAKE WAVES bend in various ways in response to the low-velocity zone, depending on whether the earthquake occurs at the surface of the earth (*top*), at the top of the low-velocity zone (*second from top*) or below the zone (*third from top*). In the first two cases, refraction of waves by the zone creates a "shadow zone" (*hatching*), where direct waves from the earthquake do not appear. The bottom diagram shows how waves from an earthquake at the surface would travel if there were no low-velocity zone to bend them down. The graphs at left show changes in the velocity of earthquake waves with increasing depth.

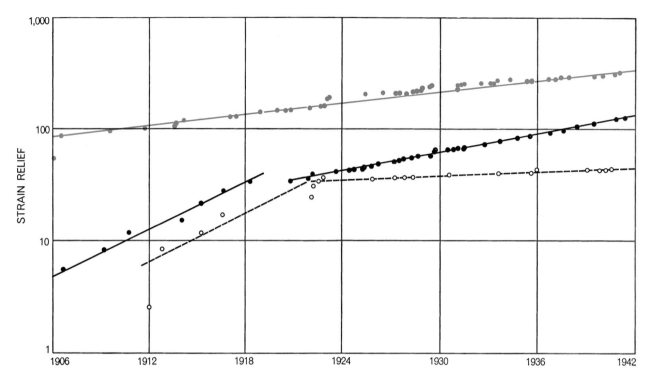

SEQUENCE OF EARTHQUAKES between 1906 and 1942 along west coast of South America falls into three groups: those occurring down to a depth of 70 kilometers (*colored dots, colored line*), those from 70 to 250 or 300 kilometers (*black dots, black line*) and those from 300 to 600 kilometers (*open dots, broken line*). The vertical scale shows relative amount of strain relieved by the quakes. Break in two lower groups around 1921 shows that they are mechanically coupled and are quite separate from the upper group.

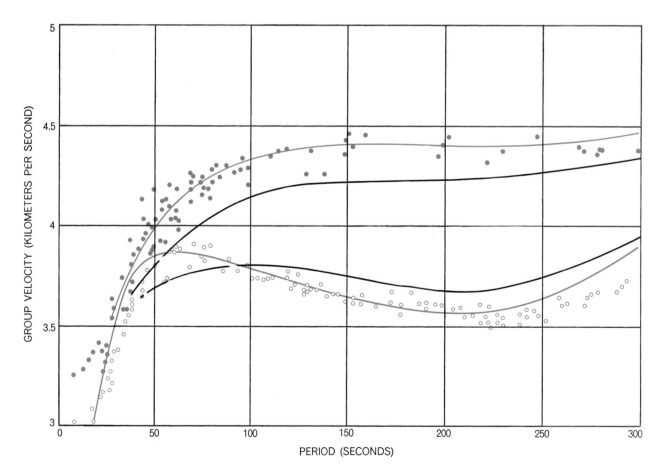

SURFACE-WAVE DATA reflect existence of low-velocity zone. The two types of dot represent actual observations of velocity of two kinds of surface waves plotted against wave period or length. Theoretical curves for an earth with a low-velocity zone (*colored lines*) fit observational data far more closely than do theoretical curves for an earth without a low-velocity zone (*black lines*).

depth that depends on their wavelength; the longer the wave, the deeper it feels [see "Long Earthquake Waves," by Jack Oliver; SCIENTIFIC AMERICAN Offprint 827]. Since in general elasticity increases with depth, longer waves travel faster than shorter ones, and waves that start out together are dispersed, or spread out. Detailed analyses of the dispersion patterns show that elasticity does not increase continuously with depth but falls off in the region of the low-velocity layer.

Body waves, which pass through the deep interior, provide only a point-by-point sampling of the outer regions of the earth. Surface waves, on the other hand, contain information about these regions over their entire path. Recent studies of surface waves in our laboratory at the California Institute of Technology and at Columbia University have demonstrated for the first time that the low-velocity layer is present below the oceans as well as below the continents. Some of the waves used in the analysis had traveled around the earth as many as seven times. They indicate that the layer is in fact a world-wide phenomenon. Comparison of oceanic and continental paths shows that the waves are slowed more under the oceans. Evidently the geological differences between ocean basins and land masses are not limited to the crust but extend several hundred kilometers into the mantle.

Conclusive proof of the world-wide extent of the low-velocity layer came from the great Chilean earthquake of May 22, 1960. It was so violent that it set the earth as a whole into vibration, making it "ring" like a bell. The tone of a bell—that is, the frequencies at which it vibrates—depends on its elastic properties; a steel and a bronze bell emit different sounds. From records of the free vibrations following a big earthquake it is possible, with enormous mathematical labor, to deduce the elastic structure of the earth. The labor has been performed. It shows that the low-velocity zone is necessary to account for the observed frequencies.

In an attempt to construct a model of the earth that fits the current seismic data, I have been obliged to conclude that the low-velocity zone transmits the horizontal and vertical vibrations in shear waves at different speeds. A crystalline material in which the crystal grains were aligned in one direction would behave this way. One mechanism that could bring about such an alignment is a flow of the material. Others are directional heat flow and differential stress.

In addition to the purely seismic data, several other phenomena attest to a lowered rigidity in the material near the top of the mantle. Variations in atmospheric pressure cause measurable deflections of the earth's surface. The amount of deflection is much greater than it would be if the crust and mantle had the same strength. By assuming a weak layer in the upper mantle the observations can be explained quite well. Moreover, most earthquakes originate in the first 60 kilo-

meters below the surface, at an average depth of 25 kilometers. At a depth of more than 60 kilometers the number falls abruptly, indicating a sudden drop in the strength of the rock.

From 60 kilometers down the frequency of earthquakes decreases steadily, dying away to zero at about 700 kilometers. This distribution implies that the rock becomes less brittle all the way from 60 to 700 kilometers and that it does not regain its strength at any deeper

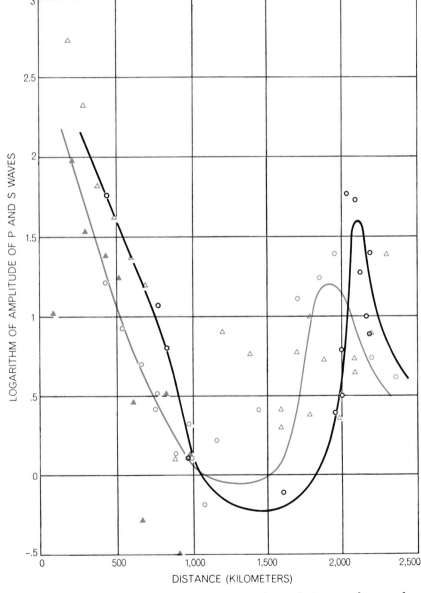

— EARTHQUAKE P WAVES, AVERAGE
○ EARTHQUAKE P WAVES RECORDED IN PASADENA
△ TWO NEVADA NUCLEAR EXPLOSIONS (P WAVES)
▲ NEW MEXICO NUCLEAR EXPLOSION (P WAVES)
— EARTHQUAKE S WAVES, AVERAGE
○ EARTHQUAKE S WAVES RECORDED IN PASADENA

SHARP DROP IN AMPLITUDE of earthquake and nuclear-explosion waves between about 100 kilometers and 1,000 kilometers from the event is caused by low-velocity zone. The two curves of averages for earthquake P and S waves were drawn by Beno Gutenberg on basis of data from many earthquakes and observatories and represent world-wide averages.

level. The picture agrees with a nomenclature first proposed in 1914 by the U.S. geologist Joseph Barrell. He spoke of an upper, rigid "lithosphere" (from the Greek word *lithos*, meaning stone) and a lower, more plastic "asthenosphere" (from the Greek word *asthenes*, meaning weak). Barrell placed the boundary between the two at a depth of 100 kilometers. Now it appears to be not a sharp boundary but a transition zone starting at some 60 kilometers.

The concept of strength and weakness in the foregoing discussion applies to the time in which stresses build up to cause earthquakes. Viewed on this temporal scale the mantle undergoes a transition from a brittle to a plastic state at about 60 kilometers and thereafter increases in plasticity. On the much shorter time scale of earthquake-wave vibrations, however, the material reverts to a stronger, or more elastic, condition at a depth of more than 250 kilometers. The decrease in velocity at the top of the mantle is gradual; it is not yet clear whether the base of the low-velocity zone is characterized by a gradual or an abrupt increase in velocity.

Almost certainly the short-term properties that set apart the low-velocity layer are determined by the temperature and pressure of the mantle in relation to its melting point at different depths. In general the elasticity of any material decreases as its temperature approaches the melting point. But an increase in pressure raises the melting point and elasticity. Below the surface of the earth both temperature and pressure increase with depth, and so the two have opposing effects on the proximity to the melting point as well as on the elastic strength of rock. Presumably at a depth of about 60 kilometers temperature takes the upper hand and the rock begins to approach its melting point, growing weaker as the depth increases. This trend continues down to some 200 kilometers, where it reverses. Then pressure raises the melting point faster than the temperature increases and the material becomes more elastic (until the liquid outer core is reached). A few laboratory experiments on rock under high temperature and pressure seem to confirm this picture. Extrapolating the rather scanty data indicates a very low strength at a

depth of somewhat more than 100 kilometers.

Hugo Benioff of the California Institute of Technology has discovered a remarkable indication of discontinuity at the level of the top of the low-velocity zone. In studying a large number of earthquakes in the Pacific Ocean earthquake belt he was able to connect certain sequences of earthquakes to single fault structures. One sequence that occurred in South America between 1906 and 1942 delineates a great fault off the west coast of the continent. The fault is some 4,500 kilometers long and goes down 600 kilometers—a tenth of the distance to the center of the earth. The earthquakes related to the fault fall naturally into three groups: (1) those shallower than 70 kilometers, (2) those from 70 to 250 or 300 kilometers and (3) those from 300 to 600 kilometers [*see top illustration on page 32*]. Analysis of the earth motions in the quakes showed a marked similarity between the intermediate and deep groups but no resemblance of these to the shallower group. In particular the motions of the two deeper groups changed suddenly, and in the

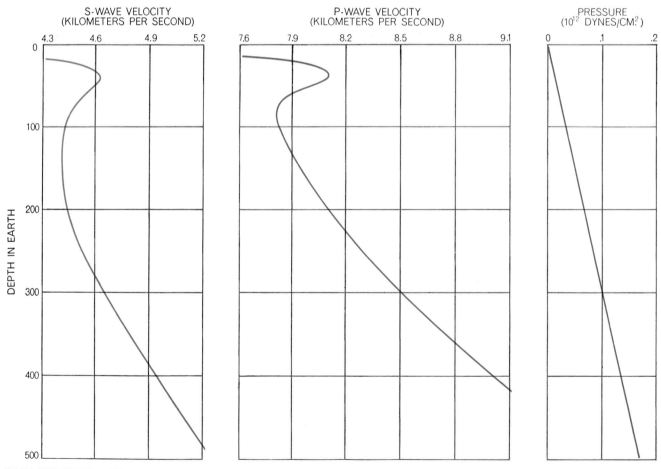

DATA FOR UPPER MANTLE reflect existence of plastic or low-velocity zone. The seismic-wave velocities and the number of earth-quakes are the only curves made from direct measurements in the earth. Pressure is derived directly from depth. Temperature curve

same way, in 1921. There was no corresponding change in the shallower earthquakes. Evidently there is some mechanical coupling between the lower layers, but these are sharply decoupled from the region above 70 kilometers. Other areas of the circum-Pacific tectonic belt show similar phenomena.

When the earthquake foci are plotted in three dimensions, those down to 250 kilometers fall in a plane about 900 kilometers wide, dipping about 33 degrees under the continents with respect to the surface of the earth. The deep earthquakes, on the other hand, are on a plane tilted at 60 degrees. Thus, although they are mechanically connected, the intermediate and deep layers are spatially discontinuous. The dimensions and location of the intermediate layer correspond closely to those of the low-velocity zone.

An interesting clue to the state of the material in the upper mantle was furnished by the Soviet volcanologist G. S. Gorshkov in 1957. He found that shear waves from Japanese earthquakes do not reach the Kamchatka Peninsula when their paths cross the volcanic belt between Japan and the peninsula. Gorshkov concluded that there must be pockets of liquid magma at a depth of 55 kilometers that absorb the waves. Apparently in certain regions the temperature not only approaches the melting point but even exceeds it. Many seismologists have remarked on the fact that the average wavelength of shear waves is many times longer than that of compressional waves. The observation could be accounted for by a weak, perhaps partially molten, layer that absorbs the shorter S waves more than the longer S waves.

Volcanoes are concentrated in parts of the world where earthquakes are most common, and the earthquakes actually associated with volcanism mostly originate at depths between 60 and 200 kilometers. This suggests that volcanoes are connected with disturbances in the region of the low-velocity zone. Therefore the distribution of volcanoes constitutes direct evidence for the temperature–melting point relation inferred from laboratory measurements and suggests that the low-velocity layer may be the source of primary basaltic magma.

Volcanism and the postglacial uplift of the crust constitute the only dynamic, as opposed to static, geological "experiments." Both indicate fluidity, and some degree of actual flow, in the material below the crust. Moreover, they are consistent with the idea of a layer of maximum plasticity in the upper mantle.

Almost all present theories of isostasy and tectonics, including those concerned with mountain building, faulting and the possible drifting of the continents, focus attention on the Mohorovičić discontinuity, which divides the crust of the earth from the mantle. If the picture I have tried to outline in this article is correct, the important discontinuity is farther down, at the ill-defined boundary of the rigid lithosphere and the weaker asthenosphere. Most of the activity responsible for the broad-scale features of the earth's surface probably takes place in a low-velocity or plastic layer at the top of the asthenosphere, extending roughly from 60 to 250 kilometers in depth. In particular the existence of such a plastic layer makes the idea of continental drift much more plausible than it has seemed heretofore.

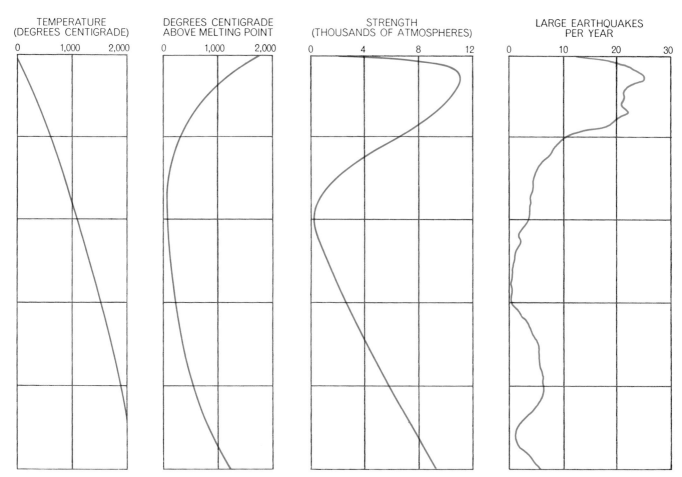

is based largely on theoretical considerations. Temperature above melting point is calculated from pressure and temperature curves.

Strength, down to 50-kilometer depth, is derived indirectly from laboratory measurements; below that it is an extrapolation.

II

CONTINENTAL DRIFT, SEA-FLOOR SPREADING, AND PLATE TECTONICS

II

CONTINENTAL DRIFT, SEA-FLOOR SPREADING, AND PLATE TECTONICS

INTRODUCTION

The articles in the preceding section are important because they laid the foundations for the later acceptance of the hypothesis of continental drift, although four of them were published prior to any revival of interest in that theory. The revival, which began in 1956, arose from two quite different causes: In that year, a group of geophysicists led by P. M. S. Blackett, Sir Edward Bullard, J. Hospers, and S. K. Runcorn showed that the scattered directions of magnetization of old rocks could be assembled into a consistent pattern, were it assumed that the continents had moved relative to the magnetic poles and to one another. In the same year, Maurice Ewing and Bruce C. Heezen suggested that a system of midocean ridges extended continuously for 40,000 miles through all the world's oceans. Evidence gathered on oceanographic expeditions soon proved the truth of this proposal. When it was shown that by far the greatest mountain system on earth lay at the bottom of the sea, the old theories were seen to be seriously deficient: existing theories of tectonics had not predicted the existence of such a system, nor could they explain its nature once it had been discovered.

In 1960, H. H. Hess made a proposal that did more than anything else to solve this puzzle. He suggested that the sea floors crack open along the crest of the midocean ridges, and that new sea floor forms there and spreads apart on either side of the crest. Wegener had held that each continent is independently propelled, so that it behaves like a ship plowing through a yielding ocean floor, whereas Hess proposed that the continents do not move as self-contained entities but as rafts frozen into, and moving with, a sea floor as rigid as they are. Wegener held that, as continents move, new sea floor forms along their retreating shores: no such activity has actually been observed, however, which is consistent with Hess's proposal that new floor is generated instead at midocean ridges. Robert S. Dietz named this process sea-floor spreading. Coupled with it was the idea that, where continents and sea floor come together, the sea floor is absorbed beneath zones of deep ocean trenches and young mountains. These were subsequently named subduction zones.

This proposal at once satisfied the geophysicists (who had objected to the notion of the ocean floors being plastic), it showed that the alpine geologists had been correct in postulating extensive shortening of strata across young mountain ranges, and it explained the nature of midocean ridges and deep ocean trenches. In a word, it provided the basis upon which to build a new and satisfying hypothesis of continental drift. The five papers in this section show how elegantly the new concept can explain many complex and previously mysterious features of the ocean floor.

The first paper in this section, "Continental Drift" by J. Tuzo Wilson, was published in 1963. It describes the spreading of the ocean floors away from the midocean ridges and attributes the motion to convection currents in the mantle. It extends the ideas of Revelle and Fisher (see their article in Section I) by showing that, as fast as new sea floor is generated along midocean ridges, older sea-floor material is being pushed down into trenches and under island arcs, where it is resorbed by the mantle.

The paper also notes that oceanic islands tend to increase in age away from midocean ridges, and that at some points along those ridges, such as Iceland and Tristan da Cunha, an excessive outpouring of lava has built active volcanic islands that are connected to the shores of adjacent continents by lateral ridges. If the "hot spots" under these islands have existed since the oceans began to open, their excess lava could have built those lateral ridges. In a similar way, the Hawaiian islands—which become progressively older north-westward along the chain away from the active island of Hawaii— could have formed over a hot spot that is now beneath Hawaii as the floor of the Pacific passed slowly across it while moving northwest-ward from the East Pacific Rise towards the trenches off Japan.

Patrick M. Hurley's article "The Confirmation of Continental Drift" shows how much progress was made in the next five years. Hurley and his colleagues determined the ages of many Precambrian rocks in Africa and South America and found that the ages of such rocks along the Atlantic coast of one continent match those of the rocks along the corresponding coast of the other. The evidence is clear that the geology and the topography of the two sides of the Atlantic are well matched, both north and south of the equator. Hurley's map shows the state of investigation of the midocean ridges at the time the article was published—many great fracture zones that offset the Mid-Atlantic Ridge had been discovered by that time—and the article gives an early account of the magnetic evidence for sea-floor spread-ing. A comparison of Hurley's reconstruction of the Indian Ocean with that in the preceding paper shows considerable differences, which emphasizes that the Indian Ocean is more complex than others and, despite the advances of the preceding five years, much uncer-tainity about particulars remained in 1968.

In the meantime, Victor Vacquier, Arthur D. Raff, and Ronald G. Mason discovered a regular pattern of magnetic anomalies off the west coast of the United States extending for hundreds of miles as elongated, parallel strips. At intervals, large topographic distur-bances (fracture zones) offset the strips by hundreds of miles. The fracture zones appear to be the traces of great strike-slip faults. A curious feature of these faults is that they all stop at coasts and do not extend into continents.

The third paper in this section, "Sea-Floor Spreading" by J. R. Heirtzler, shows how F. J. Vine and D. H. Matthews interpreted a similar pattern in the Indian Ocean. Vine and Matthews suggested that the geomagnetic field reverses at intervals of a few tens—or hundreds—of thousands of years. During any one period, the lava forming at the crest of the spreading midocean ridge is magnetized uniformly in one direction, thus forming one strip in the anomaly pattern. When the earth's field reverses, this strip is split along the crest of the ridge; as new lava wells up, it forms a new reversed anomaly within the previous one, the two halves of which move out equally on either side. With successive reversals, a symmetrical pattern of strips is built up parallel with the crest. If the times of these reversals can be ascertained, the rate of spreading can be calculated.

J. R. Heirtzler and his colleagues at Columbia University's Lamont Geological Observatory (the name has recently been changed to Lamont-Doherty Geological Observatory) had been working along similar lines, and had investigated the Reykjanes Ridge, a part of the Mid-Atlantic Ridge south of Iceland. They had reached similar views. A great body of magnetic observation was available at the observatory, and Heirtzler's paper shows how the interpretation of this data provided a time scale of 171 reversals in the past 76 million years; the demonstration that this same pattern marks all the earth's ocean floors gave great support to the concept of sea-floor spreading. This paper also introduces the concept of plate tectonics, a corollary of sea-floor spreading proposed by W. J. Morgan: because the ocean floors are rigid, motions in the lithosphere are confined to the boundaries between a few gigantic plates. It is the rubbing together of these plates that produces earthquakes; a plot of earthquake foci, therefore, will delineate plate boundaries.

H. W. Menard, who has worked for many years in the Pacific Ocean basin, is the author of the article "The Deep-Ocean Floor." This paper describes in some detail the processes that take place at the crest of the midocean ridge as the plates grow. Menard discusses the origin of fracture zones and what happens to them when the direction of plate motion changes, as it is bound to do as growth and subduction change the shapes and sizes of the plates. He also shows how the sea floor sinks as it moves away from the crest and how that influences the growth and characteristics of volcanic islands and submarine volcanoes and the seamounts that are formed from them.

The fifth paper, "The Origin of the Oceans," was written by Sir Edward Bullard, who directed the work of Vine and Matthews at the University of Cambridge. This paper gives an account of another of his contributions—one that he made with J. E. Everett and A. Gilbert Smith by using a computer to determine the best fit between opposite coasts of continents. This paper enlarges upon the concept of plate tectonics by mapping the way in which the lithosphere is broken into six large plates and about a dozen smaller ones. These plates grow and move apart along midocean ridges. They come together to form subduction zones, so that one is forced below another and swallowed up along trenches and young mountains. Where plates slide past one another, great strike-slip fault systems are formed. The San Andreas fault in California is part of such a system. The article emphasizes that the simplicity of the geology of the main ocean basins, which are produced where plates move apart, is in striking contrast to the complexity produced where plates collide.

The five papers in this section show how the ideas of sea-floor spreading and plate tectonics evolved. Since they were written, the principles have remained the same.

6

CONTINENTAL DRIFT

J. TUZO WILSON
April 1963

Geology has reconstructed with great success the events that lie behind the present appearance of much of the earth's landscape. It has explained many of the observed features, such as folded mountains, fractures in the crust and marine deposits high on the surface of continents. Unfortunately, when it comes to fundamental processes—those that formed the continents and ocean basins, that set the major periods of mountain-building in motion, that began and ended the ice ages—geology has been less successful. On these questions there is no agreement, in spite of much speculation. The range of opinion divides most sharply between the position that the earth has been rigid throughout its history, with fixed ocean basins and continents, and the idea that the earth is slightly plastic, with the continents slowly drifting over its surface, fracturing and reuniting and perhaps growing in the process. Whereas the first of these ideas has been more widely accepted, interest in continental drift is currently on the rise. In this article I shall explore the reasons why.

The subject is large and full of pitfalls. The reader should be warned that I am not presenting an accepted or even a complete theory but one man's view of fragments of a subject to which many are contributing and about which ideas are rapidly changing and developing. If it is conceded that much of this is speculation, then it should also be added that many of the accepted ideas have in fact been speculations also.

In the past several different theories of continental drift have been advanced and each has been shown to be wrong in some respects. Until it is indisputably established that such movements in the earth's crust are impossible, however, a multitude of theories of continental drift remain to be considered. Although there

is only one pattern for fixed continents and a rigid earth, many patterns of continental migration are conceivable.

The traditional rigid-earth theory holds that the earth, once hot, is now cooling, that it became rigid at an early date and that the contraction attendant on the cooling process creates compressive forces that, at intervals, squeeze up mountains along the weak margins of continents or in deep basins filled with soft sediments. This view, first suggested by Isaac Newton, was quantitatively established during the 19th century to suit ideas then prevailing. It was found that an initially hot, molten earth would cool to its present temperature in about 100 million years and that, in so doing, its circumference would contract by at least tens and perhaps hundreds of miles. The irregular shape and distribution of continents presented a puzzle but, setting this aside, it was thought that the granitic blocks of the continents had differentiated from the rest of the crustal rock and had frozen in place at the close of the first, fluid chapter of the earth's history. Since then they had been modified *in situ*, without migrating.

This hypothesis, in its essentials, still has many adherents. They include most geologists, with notable exceptions among those who work around the margins of the southern continents. The validity of the underlying physical theory is defended by some physicists. On the other hand, a number of formidable objections have been raised by those who have studied radioactivity, ancient climates, terrestrial magnetism and, most recently, submarine geology. Many biologists have also thought that, although the evolution and migration of later forms of life—particularly since the advent of mammals—could be satisfactorily traced on the existing pattern of continents, the distribution of earlier forms required either land bridges across the

oceans—the origin and disappearance of which are difficult to explain—or a different arrangement of the continents.

The discovery of radioactivity altered the original concept of the contraction theory without absolutely invalidating it. In the first place, the age of the earth could be reliably determined from knowledge of the rate at which the unstable isotopes of various elements decay and by measurement of the ratios of daughter to parent isotopes present in the rocks. These studies showed the earth to be much older than had been imagined, perhaps 4.5 billion years old. Dating of the rocks indicated that the continents are zoned and have apparently grown by accretion over the ages. Finally, it was found that the decay of uranium, thorium and one isotope of potassium generates a large but unknown supply of heat that must have slowed, although it did not necessarily stop, the cooling of the earth.

The rigid earth now appeared to be less rigid. It became possible to explain the knowledge, already a century old, that great continental ice sheets had depressed the earth's crust, just as the loads of ice that cover Greenland and Antarctica depress the crust in those regions today. Observation showed that central Scandinavia and northern Canada, which had been covered with glacial ice until it melted 11,000 years ago, were still rising at the rate of about a centimeter a year. Calculations of the viscosity of the interior based on these studies led to the realization that the earth as a whole behaves as though a cool and brittle upper layer, perhaps 100 kilometers thick, rests on a hot and plastic interior. All the large topographical features—continents, ocean basins, mountain ranges and even individual volcanoes—slowly seek a rough hydro-

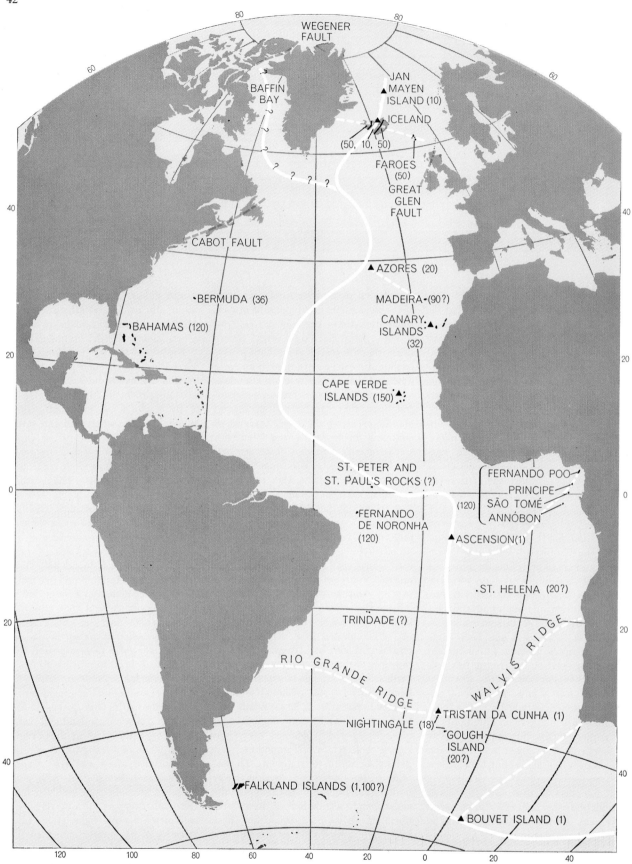

AGE OF ATLANTIC ISLANDS, as indicated by the age of the oldest rocks found on them, apparently tends to increase with increasing distance from the Mid-Atlantic Ridge. The numbers associated with the islands give these ages in millions of years. Geologists divide Iceland into three areas of different ages, the central one being the youngest. The Rio Grande and Walvis ridges are lateral ridges that may have formed as a result of the drifting apart of Africa and South America. Other lateral ridges along the Mid-Atlantic Ridge are also represented. Islands that have active volcanoes are represented by black triangles; most of these islands lie on or near the Mid-Atlantic Ridge. The extension of the ridge into Baffin Bay is postulated. Broken colored lines are faults.

static equilibrium with one another on the exterior. Precise local measurements of gravity showed that the reason some features remain higher than others is that they have deeper, lighter roots than those that are low. The continents were seen to float like great tabular icebergs on a frozen sea.

Everyone could agree that in response to vertical forces the outer crustal layer moved up and down, causing flow in the interior. The crux of the argument between the proponents of fixed and of drifting continents became the question of whether the outer crust must remain rigid under horizontal forces or whether it could respond to such forces by slow lateral movements.

Gondwanaland and "Pangaea"

Suggestions that the continents might have moved had been advanced on various grounds for centuries. The remarkable jigsaw-puzzle fit of the Atlantic coasts of Africa and South America provoked the imagination of explorers almost as soon as the continental outlines appeared opposite each other on the world map. In the late 19th century geologists of the Southern Hemisphere were moved to push the continents of that hemisphere together in one or another combination in order to explain the parallel formations they found, and by the turn of the century the Austrian geologist Eduard Suess had reassembled them all in a single giant land mass that he called Gondwanaland (after Gondwana, a key geological province in east central India).

The first comprehensive theory of continental drift was put forward by the German meteorologist Alfred Wegener in 1912. He argued that if the earth could flow vertically in response to vertical forces, it could also flow laterally. In support of a different primeval arrangement of land masses he was able to point to an astonishing number of close affinities of fossils, rocks and structures on opposite sides of the Atlantic that, he suggested, ran evenly across, like lines of print when the ragged edges of two pieces of a torn newspaper are fitted together again. According to Wegener all the continents had been joined in a single supercontinent about 200 million years ago, with the Western Hemisphere continents moved eastward and butted against the western shores of Europe and Africa and with the Southern Hemisphere continents nestled together on the southern flank of this "Pangaea." Under the action of forces

GREAT GLEN FAULT in Scotland is named for a valley resulting from erosion along the line of the fault. About 350 million years ago the northern part of Scotland was slowly moved some 60 miles to the southwest along this line (see *illustration on opposite page*).

ASPY FAULT in northern Nova Scotia is marked by several cliffs like the one seen here. The fault is part of the Cabot Fault system extending from Boston to Newfoundland (see *illustration on opposite page*) and may represent an extension of the Great Glen Fault.

associated with the rotation of the earth, the continents had broken apart, opening up the Atlantic and Indian oceans.

Between 1920 and 1930 Wegener's hypothesis excited great controversy. Physicists found the mechanism he had proposed inadequate and expressed doubt that the continents could move laterally in any case. Geologists showed that some of Wegener's suggestions for reassembling the continents into a sin-

gle continent were certainly wrong and that drift was unnecessary to explain the coincidences of geology in many areas. They could not, however, dispute the validity of most of the transatlantic connections. Indeed, more such connections have been steadily added.

It was the discovery of one of these connections that prompted my own recent inquiries into the subject of continental drift. A huge fault of great age

bisects Scotland along the Great Glen in the Caledonian Mountains. On the western side of the Atlantic, I was able to show, a string of well-known faults of the same great age connect up into another huge fault, the "Cabot Fault" extending from Boston to northern Newfoundland. These two great faults are much older than the submarine ridge and rift recently discovered on the floor of the mid-Atlantic and shown to be a

CONVECTION CURRENTS in the earth's mantle may move blocks of crustal material with different effects. Continental mountain

chains and island arcs could form where currents sink and blocks meet; mid-ocean ridges, where currents rise and blocks are torn

young formation. The two faults would be one if Wegener's reconstruction or something like it were correct. Wegener also thought that Greenland (where he died in 1930) and Ellesmere Island in the Canadian Arctic had been torn apart by a great lateral displacement along the Robeson Channel. The Geological Survey of Canada has since discovered that the Canadian coast is faulted there.

Many geologists of the Southern Hem-isphere, led by Alex. L. Du Toit of South Africa, welcomed Wegener's views. They sought to explain the mounting evidence that an ice age of 200 million years ago had spread a glacier over the now scattered continents of the Southern Hemisphere. At the same time, according to the geological record, the great coal deposits of the Northern Hemisphere were being formed in tropical forests as far north as Spitsbergen. To resolve this climatic paradox Du Toit proposed a different reconstruction of the continent. He brought the southern continents together at the South Pole and the northern coal forests toward the Equator. Later, he thought, the southern continent had broken up and its component subcontinents had drifted northward.

The compelling evidence for the existence of a Gondwanaland during the

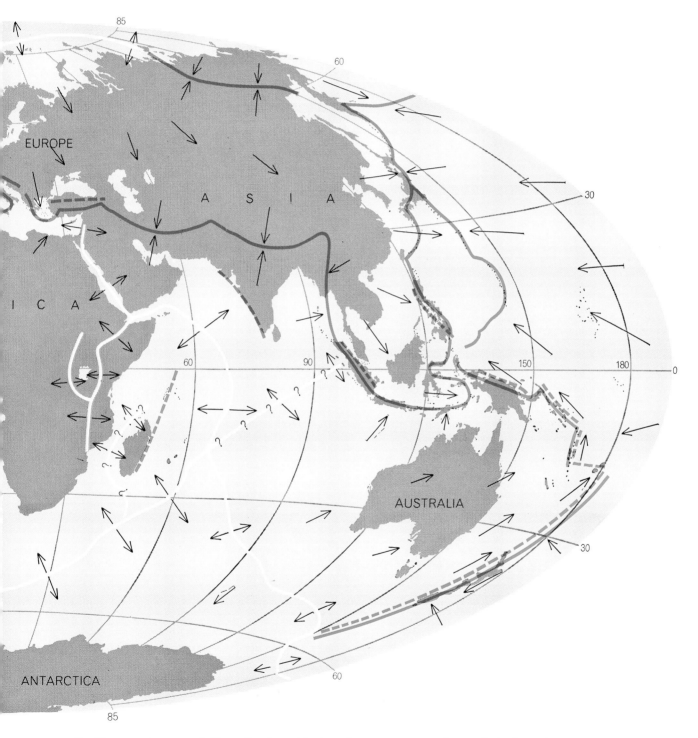

apart. On this assumption arrows indicate directions of horizontal flow of currents at the present time. Solid colored lines represent mountain chains and island arcs; heavy white lines, the worldwide system of mid-ocean ridges; and broken colored lines, faults.

Mesozoic era—the "Age of Reptiles"—has been reinforced by the findings made in Antarctica since the intensive study of that continent began in 1955. The ice-free outcrops on the continent, although few, not only show the record of the earlier ice age that gripped the rest of the land masses in the Southern Hemisphere but also bear deposits of a low-grade coal laid down in a still earlier age of verdure that covered all the same land masses with the peculiar big-leafed *Glossopteris* flora found in their coal beds as well.

Many suggestions have been made as to how to create and destroy the land bridges needed to explain the biological evidence without moving the continents. Some involve isthmuses and some involve whole continents that have subsided below the surface of the ocean. But the chemistry and density of continents and ocean floors are now known

to be so different that it seems even more difficult today to raise and lower ocean floors than it is to cause continents to migrate.

Convection in the Mantle

One of the first leads to a mechanism that would move continents came more than 30 years ago from the extension to the ocean floor of the sensitive techniques of gravimetry that had established the rule of hydrostatic equilibrium, or isostasy, ashore. The Dutch geophysicist Felix A. Vening Meinesz demonstrated that a submerged submarine would provide a sufficiently stable platform to allow the use of a gravimeter at sea. Over the abyssal trenches in the sea floor that are associated with the island arcs of Indonesia and the western side of the Pacific he found some of the largest deficiencies

in gravity ever recorded. It was clear that isostasy does not hold in the trenches. Some force at work there pulls the crust into the depths of the trenches more strongly than the pull of gravity does.

Arthur Holmes of the University of Edinburgh and D. T. Griggs, now at the University of California at Los Angeles, were stimulated by these observations to re-examine and restate in modern terms an old idea of geophysics: that the interior of the earth is in a state of extremely sluggish thermal convection, turning over the way water does when it is heated in a pan. They showed that convection currents were necessary to account in full for the transfer of heat flowing from the earth's interior through the poorly conductive material of the mantle: the region that lies between the core and the crust. The trenches, they said, mark the places where currents in

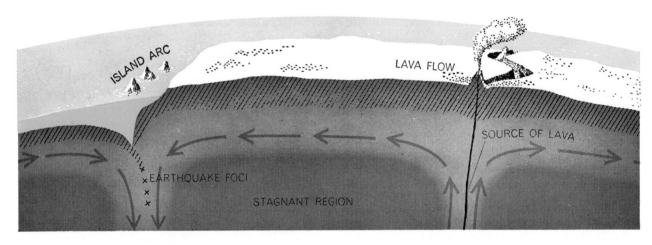

EFFECTS OF CONVECTION CURRENTS, schematized in the two illustrations on this page, provide one possible means of accounting for the formation of median ridges, lateral ridges, mountain ranges and earthquake belts. Rising and separating currents (*arrows at right*) could break the crustal rock and pull it apart; the rift would be filled by altered mantle material (as suggested by H. H. Hess of Princeton University) and lava flows, forming a median ridge. Sinking currents (*left*) could pull the ocean floor down.

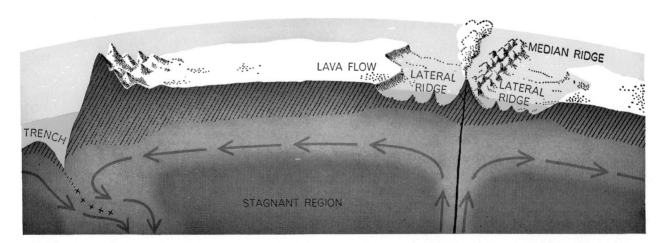

DRIFTING CONTINENT may be "piled up," where it meets sinking currents, to form mountains like those of the Andes (*left*). Since continents are lighter than the mantle material of the ocean floor, they cannot sink but tend to be pushed over sinking currents, which are marked by deep earthquakes. Active volcanoes continue to form over rising currents (*right*); but drift may carry these volcanic piles away to either side of the median ridge. Separated from their source, the inactive cones form one or two lateral ridges.

the mantle descend again into the interior of the earth, pulling down the ocean floor.

Convection currents in the mantle now play the leading role in every discussion of the large-scale and long-term processes that go on in the earth. It is true that the evidence for their existence is indirect; they flow too deep in the earth and too slowly—a few centimeters a year—for direct observation. Nonetheless their presence is supported by an increasing body of independently established evidence and by a more rigorous statement of the theory of their behavior. Recently, for example, S. K. Runcorn of Durham University has shown that to stop convection the mantle material would have to be 10,000 times more viscous than the rate of postglacial recoil indicates. It is, therefore, highly probable that convection currents are flowing in the earth.

Perhaps the strongest confirmation has come with the discovery of the regions where these currents appear to ascend toward the earth's surface. This is the major discovery of the recent period of extraordinary progress in the exploration of the ocean bottom, and it involves a feature of the earth's topography as grand in scale as the continents themselves. Across the floors of all the oceans, for a distance of 40,000 miles, there runs a continuous system of ridges. Over long stretches, as in the mid-Atlantic, the ridge is faulted and rifted under the tension of forces acting at right angles to the axis of the ridge. Measurements first undertaken by Sir Edward Bullard of the University of Cambridge show that the flow of heat is unusually great along these ridges, exceeding by two to eight times the average flow of a millionth of a calorie per square centimeter per second observed on the continents

and elsewhere on the ocean floor. Such measurements also show that the flow of heat in the trenches, as in the Acapulco Trench off the Pacific coast of Central America, falls to as little as a tenth of the average.

Most oceanographers now agree that the ridges form where convection currents rise in the earth's mantle and that the trenches are pulled down by the descent of these currents into the mantle. The possibility of lateral movement of the currents in between is supported by evidence for a slightly plastic layer—called the asthenosphere—below the brittle shell of the earth. Seismic observations show that the speed of sound in this layer suddenly becomes slower, indicating that the rock is less dense, hotter and more plastic. These observations have also yielded evidence that the asthenosphere is a few hundred kilometers thick, somewhat thicker than

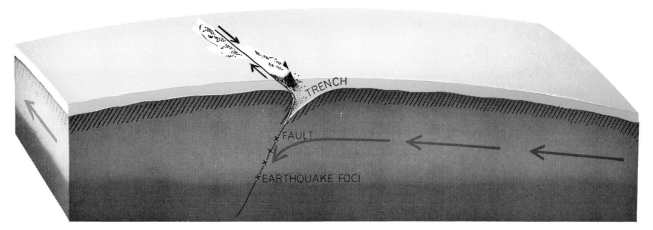

TWO CONVECTION CURRENTS perpendicular to each other suggest a mechanism for producing large horizontal faults such as the one that has offset western New Zealand 300 miles northward. The two convection currents (*arrows indicate direction*) would produce a fault. One current would be forced downward, producing a trench and earthquakes along the sloping surface. Continued flow of the second current would result in a sliding motion, or lateral displacement, along the plane of the fault, shearing the island in two.

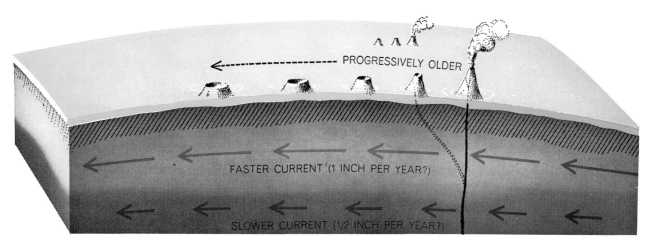

VOLCANIC-ISLAND CHAINS like the Hawaiian Islands must have originated in a process slightly different from that which formed pairs of lateral ridges. The source of lava flow does not lie on a mid-ocean ridge; it is considered that the source may be deep (100 miles or more) in the slower moving part of convection currents. The differential motion carries old volcanoes away from the source, while new volcanoes form over the source. The length of the island chain depends on how long the source has been active.

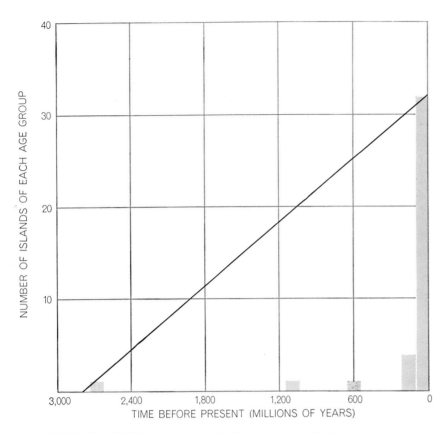

FREQUENCY DIAGRAM shows the age distribution of about 40 islands (in main ocean basins) dated older than "recent" (the number of very young islands is vastly greater). The diagonal line shows the corresponding curve for continental rock ages over equivalent areas.

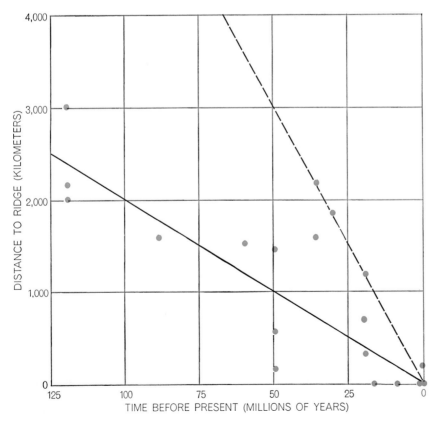

DISTANCE FROM MID-OCEAN RIDGE of some islands in Atlantic and Indian oceans is plotted against age. If all originated over the ridge, their average rate of motion has been two centimeters a year (*solid line*); maximum rate, six centimeters a year (*broken line*).

the crust, and that below it the viscosity increases again.

Here, then, is a mechanism, in harmony with physical theory and much geological and geophysical observation, that provides a means for disrupting and moving continents. It is easy to believe that where the convection currents rise and separate, the surface rocks are broken by tension and pulled apart, the rift being filled by the altered top of the mantle and by the flow of basalt lavas. In contrast to earlier theories of continental drift that required the continents to be driven through the crust like ships through a frozen sea, this mechanism conveys them passively by the lateral movement of the crust from the source of a convection current to its sink. The continents, having been built up by the accumulation of lighter and more siliceous materials brought up from below, are not dragged down at the trenches where the currents descend but pile up there in mountains. The ocean floor, being essentially altered mantle, can be carried downward; such sediments as have accumulated in the trenches descend also and, by complicated processes, may add new mountains to the continents. Since the material near the surface is chilled and brittle, it fractures, causing earthquakes until it is heated by its descent.

From the physical point of view, the convection cells in the mantle that drive these currents can assume a variety of sizes and configurations, starting up and slowing down from time to time, expanding and contracting. The flow of the currents on the world map may therefore follow a single pattern for a time, but the pattern should also change occasionally owing to changes in the output and transfer of heat from within. It is thus possible to explain the periodicity of mountain-building, the random and asymmetrical distribution of the continents and the abrupt breakup of an ancient continent.

Some geophysicists consider that isostatic processes set up by gravitational forces may suffice to cause the outer shell to fracture and to slip laterally over the plastic layer of the asthenosphere. This mechanism would not require the intervention of convection currents. Both mechanisms could explain large horizontal displacements of the crust.

Evidence from Terrestrial Magnetism

Fresh evidence that such great movements have indeed been taking place has been provided by two lines of study in

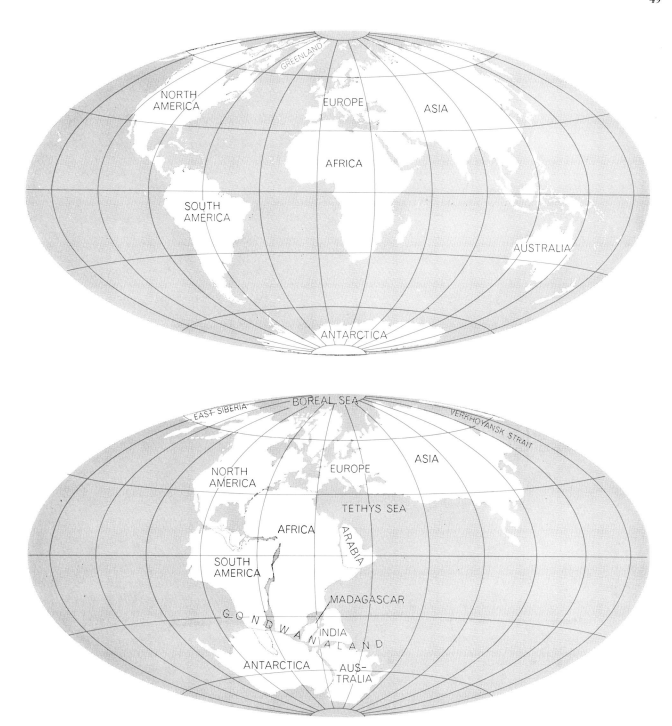

SINGLE SUPERCONTINENT, presumed to have existed some 150 million years ago, would have resembled that depicted in the map **at bottom. A present-day map appears at top. In both maps the distortion of the continents is a result of the projection employed.**

the field of terrestrial magnetism. On the one hand, surveys of the earth's magnetic field off the coast of California show a pattern of local anomalies in the ocean floor running parallel to the axis of a now inactive oceanic ridge that underlies the edge of the continent. The pattern bears a persuasive resemblance to the "photoelastic" strain patterns revealed by polarized light in plastics placed under stress. More important, the pattern shows that the ocean floor

is faulted at right angles to the axis of the ridge, with great slabs of the crust displaced laterally to the west by as much as 750 miles. These are apparently ancient and inactive fractures; now the active faults run northwesterly, as is indicated by the earthquakes along California's San Andreas Fault.

Evidence of a more general nature in favor of continental drift comes from the studies of the "remanent" magnetism of the rocks, to which Runcorn, P. M. S.

Blackett of the University of London and Emil Thellier of the University of Paris have made significant contributions. Their investigations have shown that rocks can be weakly magnetized at the time of formation—during cooling in the case of lavas and during deposition in the case of sediments—and that their polarity is aligned with the direction of the earth's magnetic field at the place and time of their formation. The present orientation of the rocks of various ages on the continents indicates that they must have been formed in different latitudes. The rocks of any one continent show consistent trends in change of orientation with age; those from other continents show different shifts. Continental drift offers the only explanation of these findings that has withstood analysis.

Some physicists and biologists are now prepared to accept continental drift, but many geologists still have no use for the hypothesis. This is to be expected. Continents are so large that much geology would be the same whether drift had occurred or not. It is the geology of the ocean floors that promises to settle the question, but the real study of that two-thirds of the

earth's surface has just begun.

The Oceanic Islands

One decisive test turns on the age of the ocean floor. If the continents have been fixed, the ocean basins should all be as old as the continents. If drift has occurred, some regions of the ocean floor should be younger than the time of drift.

A survey of the scattered and by no means complete literature on the oceanic islands conducted by our group at the University of Toronto shows that of all the islands in the main ocean basins only about 40 have rocks that have been dated older than the Recent epoch. Only three of these—Madagascar and the

Seychelles of the Indian Ocean and the Falklands of the South Atlantic—have very old rocks; all the others are less than 150 million years old. If one regards the exceptions as fragments of the nearby continents, the youth of the others suggests that either the ocean basins are young or that islands are not representative samples of the rock of the ocean floor.

Significantly, it turns out that the age of the islands in the Atlantic Ocean tends to increase with their distance from the mid-ocean ridge. In this reckoning one need not count the island arcs of the West Indies or the South Sandwich Islands, which belong to the Cordilleran system—that is, the spine of mountains running the entire length of North and South America—and so have a continental origin. At least six of the islands on the ridge or very close to it have on them active volcanoes that have had recent eruptions; the most recent was the eruption of Tristan da Cunha, which is located squarely on the ridge in the South Atlantic. Only two of the islands far from the ridge have active volcanoes. If the •hot convection currents of the mantle rise under the mid-ocean ridge, it is easy to understand why the ridge is the locus of active volcanoes and earthquakes. The increase in age with distance from the ridge suggests that if the more distant islands had a volcanic origin on the ridge, lateral movement of the ocean floor has carried them away from the ridge. Their ages and distances from the ridge indicate movement at the rate of two to six centimeters a year on the average, in keeping with the estimated velocity of the convection currents.

Of great significance in connection with the mechanism postulated here are the two lateral ridges that run east and west from Tristan da Cunha to Africa on the one hand and to South America on the other. It is reasonable to suppose that these ridges had their origin in a succession of volcanoes that erupted and grew into mountains on the site of the present volcano and were carried off east and west to form a row of progressively older, extinct and drowned volcanoes [*see illustration on page 42*]. There are no earthquakes along the lateral ridges and so they are distinctly different in character from the mid-ocean ridge. These ridges meet the continental margins at places that would fit together on the quite independent criterion of the match of their shore lines. One explanation of this coincidence is that the continents were indeed joined together and have moved apart, with the

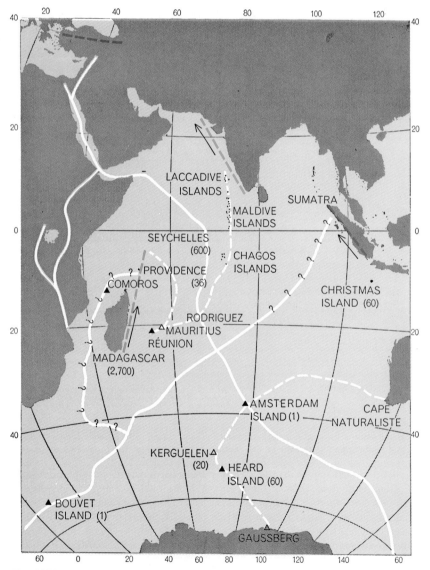

INDIAN OCEAN possibly formed as the result of four continents drifting apart. If so, four median ridges would have formed midway between continents, with pairs of lateral ridges connecting them. Heavy white lines show three known median ridges; there is evidence for one running to Sumatra. Broken white lines are lateral ridges; broken colored lines, faults; open triangles, inactive volcanoes. Numbers give ages in millions of years.

lateral ridges forming trails that record the motion. The two ridges are roughly mirror images of each other, showing that the motion was uniform on each side. Another similar pair of ridges connects Iceland—where the mid-ocean ridge comes to the surface and where the great tension rift is visible in the Icelandic Graben—to Greenland and the shelf of the European continent.

We have therefore advanced two related hypotheses: first, that where adjacent continents were once joined a median ridge should now lie between them; second, that where such continents are connected by lateral ridges they were once butted together in such a manner that points marked by the shoreward ends of these ridges coincided. If this is correct, it provides a unique method for reassembling continents that have drifted apart. One of the major troubles with theories of drift has been that the possibilities are so numerous no such precise criterion existed for putting the poorly fitting jigsaw puzzle together.

Without doubt the most severe test of this double hypothesis is presented by the Indian Ocean. Here four continents—Africa, India, Australia and Antarctica—may be assumed on geological and paleomagnetic evidence to have drifted apart. The collision of India with the Asian land mass could have thrown up the Himalaya mountains at their junction. These continents should accordingly be separated by four mid-ocean ridges. Three such ridges have

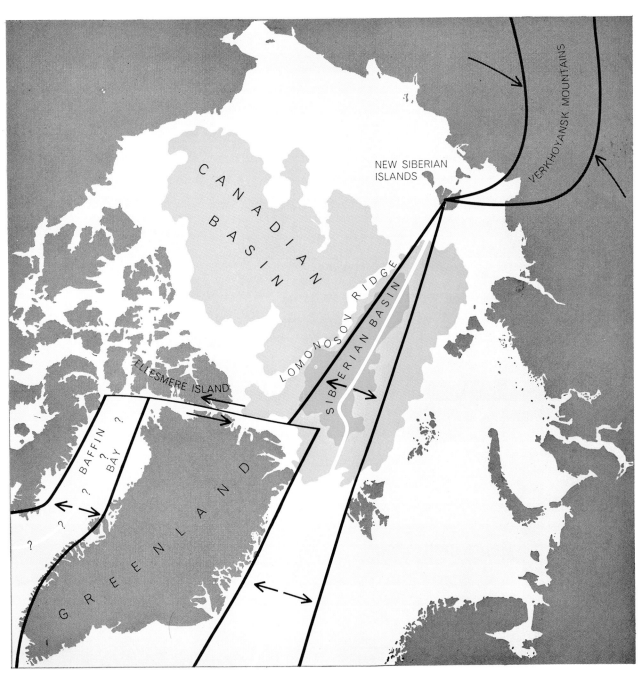

RIFTING OF SUPERCONTINENT to form the Atlantic Ocean could have produced the Verkhoyansk Mountains in eastern Siberia. As shown on this map of the Arctic, the rift spread more widely to the south. The opening of the Atlantic Ocean and Baffin Bay separated Greenland from both North America and Europe. The continents were rotated slightly about a fulcrum near the New Siberian Islands. The resulting compression and uplift would create a mountain range. Opposing arrows mark the Wegener Fault.

already been well established by surveys of the Indian Ocean, and there is evidence for the existence of the fourth. In each quadrant marked off by the ridges there is also, it happens, a lateral ridge! These submarine trails may be presumed to be records of the motion of the continents as they receded from one another. From Amsterdam Island one of these lateral ridges runs through Ker-

guelen Island to Gaussberg Mountain on the coast of Antarctica; a mirror image of this ridge runs from Amsterdam Island to Cape Naturaliste on Australia. The corresponding ridges connecting Africa and India are distorted by lateral faults running along the coasts of Madagascar and India. Thus in each quadrant there exists a lateral ridge to show how points on Madagascar, India,

Australia and Antarctica once lay close together. What is remarkable is not that there is some irregularity in the present configuration of these ridges but that the floor of the Indian Ocean should show such a symmetrical pattern.

The mid-ocean ridge separating Australia from Antarctica has been traced by Henry W. Menard of the Scripps Institution of Oceanography across the

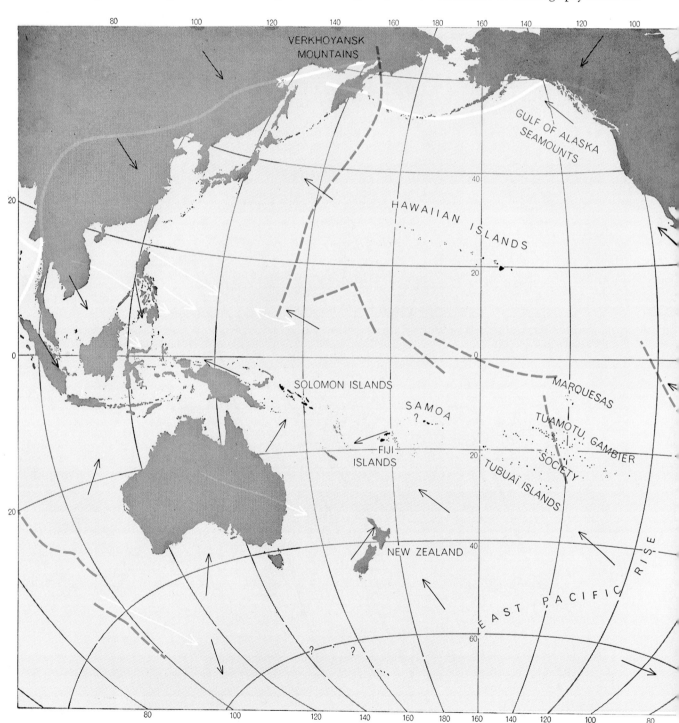

AGE OF PACIFIC ISLANDS appears to increase with increasing distance from the mid-ocean ridge. This is compatible with the idea that the eastern half of the Pacific Ocean has been spreading from the East Pacific Rise (as has been suggested by Robert S. Dietz of the U.S. Navy Electronics Laboratory). Broken colored lines represent faults; the associated arrows indicate the direction of horizontal motion, where known, along the fault. Other arrows show the probable directions of convection flow. Island arcs of the kind

eastern Pacific to connect with the great East Pacific Rise. From the topography of the Pacific floor it can be deduced that this ridge once extended through the rise marked by Cocos Island off Central America and formed the rifted ridge that moved North and South America apart. Another branch of this ridge, running across the southern latitudes, suggests the cause of the separation of South America from Antarctica. The oceanic islands in this broad region of the Pacific form lines that extend at right angles down the flanks of the East Pacific Rise; geologists long ago established that these islands grow progressively older with distance from the top of the rise. Unlike the rest of the continuous belt of mid-ocean ridges to which it is connected, the East Pacific Rise tends to run along the margins of the Pacific Ocean; it has rifted an older ocean apart rather than a continent. The floor of the western Pacific is believed to be a remnant of that older floor.

There are therefore enough connections to draw all the continents together, reversing the trends of motion indicated by the mid-ocean ridges and using the continental ends of pairs of lateral ridges as the means of matching the coast lines together. The ages of the islands and of the coastal formations suggest that about 150 million years ago, in mid-Mesozoic time, all the continents were joined in one land mass and that there was only one great ocean [*see illustration on page 49*]. The supercontinent that emerges from this reconstruction is not the same as those proposed by Wegener, Du Toit and other geologists, although all have features in common. The widespread desert conditions of the mid-Mesozoic may have been a consequence of the unusual circumstance that produced a single continent and a single ocean at that time. Since its approximate location with respect to latitude is known, along with the location of its major mountain systems, the climate in various regions might be reconstructed and compared with geological evidence.

It is not suggested that this continent was primeval. That it was in fact assembled from still older fragments is suggested by two junction lines: the ancient mountain chain of the Urals and the chain formed by the union of the Appalachian, Caledonian and Scandinavian mountains may have been thrown up in the collisions of older continental blocks. Before that there had presumably been a long history of periodic assembly and disassembly of continents and fracturing and spreading of ocean floors, as convection cells in the mantle proceeded to turn over in different configurations. At present it is impossible even to speculate about the details.

Breakup of the Supercontinent

If it can be assumed that the proposed Mesozoic continent did exist and spread apart, geology provides some guide to the history of its fragmentation. The present system of convection currents has apparently been constant in general configuration ever since the Mesozoic, but not all parts of it have been equally active all of that time. Shortly before the start of the Cretaceous period, about 120 million years ago, the continent developed a rift that opened up to form the Atlantic Ocean. The rift spread more widely in the south, with the result that the continents must have rotated slightly about a fulcrum near the New Siberian Islands [*see illustration on page 51*]. Soviet geologist have found that the compression and uplift that raised the Verkhoyansk Mountains across eastern Siberia began at about that time. To the south a continuation of the rifting separated Africa from Antarctica and spread diagonally across the Indian Ocean, opening the northeasterly rift. Africa and India were thus moved northward, away from the still intact Australian-Antarctic land mass.

It seems reasonable to suggest, particularly from the geology of the Verkhoyansk Mountains and of Iceland, that at the start of Tertiary time, about 60 million years ago, this convection system became less active and that rifting started up elsewhere. A new rift opened up along the other, northwesterly, diagonal of the Indian Ocean, separating Africa from India and Australia and separating Australia from Antarctica. With the collision of the Indian subcontinent against the southern shelf of the Asiatic land mass, the uplift of the Himalaya mountains began. The proposed succession of activity in the two main ridges of the Indian Ocean would explain why India has moved twice as far north with relation to Antarctica as Australia or Africa has and why the older northeast ridge is now a somewhat indistinct feature of the ocean floor. The younger rift in the Indian Ocean seems to have extended along the East Pacific Rise and Cocos Ridge to cross the Caribbean. A branch also passed south of South America. As these median ridges have continued to widen they have been forced by this growth to migrate northward, forming great shears or faults off the coast of Chile and through California. Indeed, a case can be made out for the idea that every mid-ocean ridge normally ends at a great fault or at a pivot point, as in the New Siberian Islands.

A few million years ago activity in this system decreased, allowing the North and South American continents to be joined by the Isthmus of Panama. The Atlantic rift now became more active again, producing renewed uplift in the

represented by Japan develop where the forces associated with such flow are directly opposed; great horizontal faults, where these forces meet at right angles.

ROBESON CHANNEL separating northwestern Greenland (*upper right*) from Ellesmere Island (*foreground*) marks the Wegener Fault. The latter was named by the author for the German meteorol-ogist who 50 years ago predicted the existence of such a fault and of a great lateral displacement along the length of the channel. Not yet fully mapped, it probably joins a known fault farther southwest.

Verkhoyansk Mountains and active volcanoes in Iceland and the five other still active volcanic islands down the Atlantic. Again the pattern of rifting in the Indian Ocean was altered. The distribution of recent earthquakes shows that the greatest activity extends along the western half of each diagonal ridge from the South Atlantic to the entrance of the Red Sea and thence by two arms along the rift valley of the Jordan River and through the African rift valleys, where the breakup of a continent has apparently begun.

The presently expanding rifts run mostly north and south or northeasterly so that dominant easterly and westerly compression of the outer crust is absorbed by overthrusting and sinking of the crust along the eastern and western sides of the "ring of fire" around the Pacific. For this reason East Asia, Oceania and the Andes are the most active regions of the world. The west-ward-driving pressure of the South Atlantic portion of the Mid-Atlantic Ridge has forced the continental block of South America against and over the downward-plunging oceanic trench along its Pacific coast. The northwest-trending currents below the Pacific floor have pulled down trenches under the eight island arcs around the western and northern Pacific from the Philippines north to the Aleutians. Even at the surface of the Pacific, the direction of the subcrustal movement is indicated by the strike of several parallel chains of volcanic islands, such as the Hawaiians, which may be thought to have risen like bubbles in a stream from the slower moving deep interior [see lower illustration on page 47]. These chains run parallel with the seismically active shearing faults that border each side of the Pacific, along the coast of North America and from Samoa to the Philippines. The compression exerted by the mid-ocean ridge through the southern seas is absorbed, with less seismic activity, along a line from New Zealand, through Indonesia and the Himalaya highlands to the European Alps. In all cases, the angle at which the loci of deep-focus earthquakes dip into the earth seems to follow the direction of subsurface flow—eastward and downward, for example under the Pacific coast of South America; westward and downward under the island arcs on the opposite side of the Pacific.

The theory I have outlined may be highly speculative, but it is indicative of current trends in thought about the earth's behavior. The older theories of the earth's history and behavior have proved inadequate to meet the new findings, particularly those from studies of terrestrial magnetism and oceanography. In favor of the specific details suggested here is the fact that they fit observations and are precise enough to be tested.

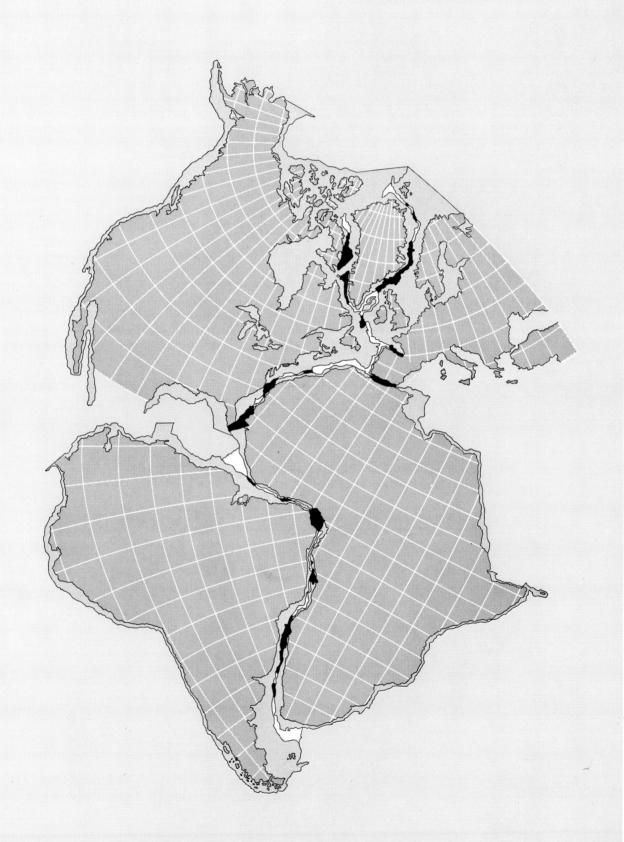

7

THE CONFIRMATION OF CONTINENTAL DRIFT

PATRICK M. HURLEY

April 1968

As recently as five years ago the hypothesis that the continents had drifted apart was regarded with considerable skepticism, particularly among American investigators. Since then, as a result of a variety of new findings, the hypothesis has gained so much support that its critics may now be said to be on the defensive. The slow acceptance of what is actually a very old idea provides a good example of the intensive scrutiny to which scientific theories are subjected, particularly in the earth sciences, where the evidence is often conflicting and where experimental demonstrations are usually not possible.

As long ago as 1620 Francis Bacon discussed the possibility that the Western Hemisphere had once been joined to Europe and Africa. In 1668 P. Placet wrote an imaginative memoir titled *La corruption du grand et du petit monde, où il est montré que devant le déluge, l'Amérique n'était point séparée des autres parties du monde* ("The corruption of the great and little world, where it is shown that before the deluge, America was not separated from the other parts of the world"). Some 200 years later Antonio Snider was struck by the similarities between American and European fossil plants of the Carboniferous period (about 300 million years ago) and proposed that all the continents were once part of a single land mass. His work of 1858 was called *La Création et Ses Mys-*

tères Dévoilés ("The Creation and Its Mysteries Revealed").

By the end of the 19th century geology had come seriously into the discussion. At that time the Austrian geologist Eduard Suess had noted such a close correspondence of geological formations in the lands of the Southern Hemisphere that he fitted them into a single continent he called Gondwanaland. (The name comes from Gondwana, a key geological province in east central India.) In 1908 F. B. Taylor of the U.S. and in 1910 Alfred L. Wegener of Germany independently suggested mechanisms that could account for large lateral displacements of the earth's crust and thus show how continents might be driven apart. Wegener's work became the center of a debate that has lasted to the present day.

Wegener advanced a remarkable number of detailed correlations, drawn from geology and paleontology, indicating a common historical record on the two sides of the Atlantic Ocean. He proposed that all the continents were joined in a single vast land mass before the start of the Mesozoic era (about 200 million years ago). Wegener called this supercontinent Pangaea. Today the evidence favors the concept of two large land masses: Gondwanaland in the Southern Hemisphere and Laurasia in the Northern.

In the Southern Hemisphere an additional correlation was found in a succession of glaciations that took place in the Permian and Carboniferous periods. These glaciations left a distinctive record in the southern parts of South America, Africa, Australia, in peninsular India and Madagascar and, as has been discovered recently, in Antarctica. The evidence of glaciations is compelling. Beds of tillite—old, consolidated glacial rubble—have been studied in known glaciated regions and are unquestioned

evidence of the action of deep ice cover. In addition many of the tillites rest on typically glaciated surfaces of hard crystalline rock, planed flat and grooved by the rock-filled ice moving over them.

This kind of evidence has been found throughout the Southern Hemisphere. In all regions the tillites are found not only in the same geological periods but also in a sequence of horizontal beds bearing fossils of identical plant species. This sequence, including the geological periods from the Devonian to the Triassic, is called the Gondwana succession. The best correlations are apparent in the Permocarboniferous beds, where two distinctive plant genera, *Glossopteris* and *Gangamopteris,* reached their peak of development. These plants were so abundant that they gave rise to the Carboniferous coal measures, which are commonly interbedded in the Gondwana succession [*see top illustration on next two pages*].

The South African geologist Alex L. du Toit and others have sought out and mapped these Gondwana sequences so diligently that today they provide the strongest evidence not only that these continental areas were joined in the past but also that they once wandered over or close to the South Pole. It is inconceivable that the complex speciation of the Gondwana plants could have evolved in the separate land masses we see today. It takes only a narrow strip of water, a few tens of miles wide at the most, to stop the spread of a diversified plant regime. The Gondwana land mass was apparently a single unit until the Mesozoic era, when it broke into separate parts. Thereafter evolution proceeded on divergent paths, leading to the biological diversity we observe today on the different continental units.

Wegener and Du Toit published their work in the 1920's and 1930's. The de-

FIT OF CONTINENTS (*opposite page*) was optimized and error-tested on a computer by Sir Edward Bullard, J. E. Everett and A. G. Smith of the University of Cambridge. Over most of the boundary the average mismatch is no more than a degree. The fit was made along the continental slope (*light gray*) at the 500-fathom contour line. The regions where land masses, including the shelf, overlap are black; gaps are white.

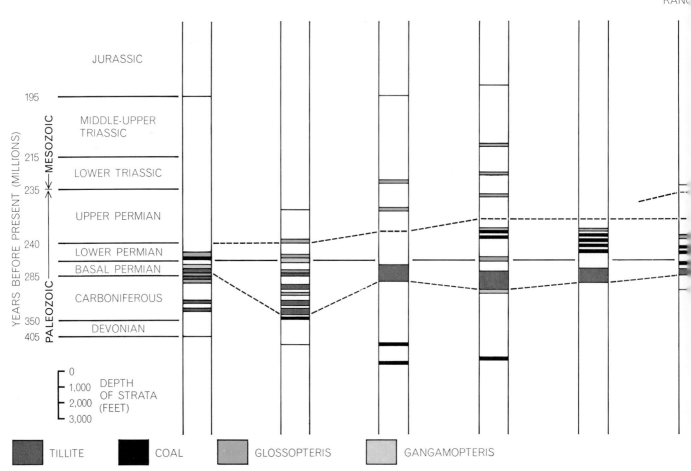

GEOLOGICAL PERIOD — ARGENTINA — BRAZIL — FALKLAND ISLANDS — SOUTH AFRICA — ANTARCTICA HORLICK MOUNTAINS — ANTARC QUEE ALEXAN RAN

JURASSIC
MIDDLE-UPPER TRIASSIC
LOWER TRIASSIC
UPPER PERMIAN
LOWER PERMIAN
BASAL PERMIAN
CARBONIFEROUS
DEVONIAN

MESOZOIC
PALEOZOIC

YEARS BEFORE PRESENT (MILLIONS)

195
215
235
240
285
350
405

0
1,000 DEPTH
2,000 OF STRATA
3,000 (FEET)

TILLITE   COAL   GLOSSOPTERIS   GANGAMOPTERIS

GONDWANA SUCCESSION is the name given to a late Paleozoic succession of land deposits found in South America, Africa, Antarctica, India and Australia. The succession contains beds of tillite (glacial rubble), coal deposits and a diversity of plants arranged in such a way that perhaps 200 million years ago the different areas must have been a single land mass known as Gondwanaland, or at

bate for and against drift became polarized largely between geologists of the Southern Hemisphere and the leaders of geophysical thought in the Western Hemisphere. Eminent geophysicists such as Sir Harold Jeffreys of the University of Cambridge voiced strong opposition to the hypothesis on the grounds that the earth's crust and its underlying mantle were too rigid to permit such large motions, considering the limited energy thought to be available.

Not all felt this way, however. In the late 1930's the Dutch geophysicist F. A. Vening Meinesz proposed that thermal convection in the earth's mantle could provide the mechanism. His ideas were supported by his gravity surveys over the deep-sea trenches and the adjacent island arcs of the western Pacific. The results implied that some force was maintaining the irregular shape of the earth's surface against its natural tendency to flatten out. Presumably the force was somehow related to thermal convection. Arthur Holmes of the University of

Edinburgh added his weight to the argument in favor of the hypothesis, and he was followed by S. W. Carey of Tasmania, Sir Edward Bullard and S. K. Runcorn of Britain, L. C. King of South Africa, J. Tuzo Wilson of Canada and others [see the article "Continental Drift," by J. Tuzo Wilson, beginning on page 41]. The historical and dynamical characteristics of the earth now engaged the attention of many more geophysicists, and today the interplay of all branches of geology and geophysics generates the excitement of a new frontier area.

Continents and Oceans

Although the general nature of the earth's crust is familiar to most readers of *Scientific American*, it is worth reviewing and summarizing some of its major features while asking: How do these features look in the context of continental drift? The earth's topography has two principal levels: the level of the

continental surface and the level of the oceanic plains. The elevations in between represent only a small fraction of the earth's total surface area. What maintains these levels? Left alone for billions of years, they should reach equilibrium at an average elevation below the present sea level, so that the earth would be covered with water. Instead we see sharp continental edges, new mountain belts, deep trenches in the oceans—in short, a topography that appears to have been regularly rejuvenated.

The continental areas are a mosaic of blocks that are roughly 1,000 kilometers across and have ages ranging from about 3,000 million years to a few tens of millions. In Africa there appear to be several ancient nuclear areas, or cratons, surrounded by belts of younger rocks. Most of the younger belts have an age of 600 million years or less, contrasting sharply with an age of 2,000 million to 3,000 million years for the cratons.

A closer look at the younger belts tells us that although much of the material is

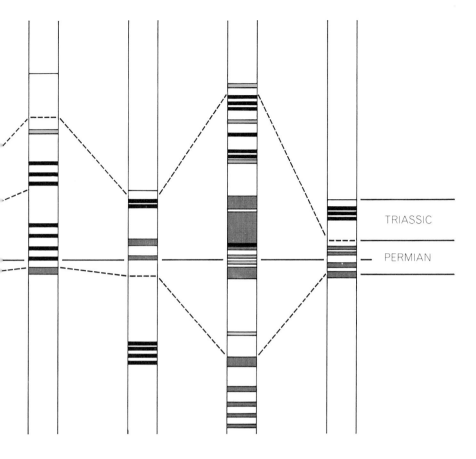

PENINSULAR
INDIA

WESTERN
AUSTRALIA

EASTERN
AUSTRALIA

TASMANIA

TRIASSIC

PERMIAN

the very least a closely associated mass connected by land bridges. Only two of several major plant genera are plotted here: *Glossopteris* and *Gangamopteris*. The depths of the various deposits have been arbitrarily aligned between the lower and the basal Permian.

continental platform. Geological mapping, however, reveals the belt structure clearly. A closer look at the cratons shows us that they too have the structure of preexisting mountain belts that have been carved into segments, with the younger material always cutting across the older structural pattern.

We see the process in action today. Our young mountain belts have not been eroded to sea level but show high elevations that are clearly apparent; we do not need geological surveys to observe them. It is only when we see the global distribution of these mountain belts on land areas, together with the distribution of rifts and their associated ridges under the oceans, that we begin to perceive the possibility that vast motions of the earth's surface may be their cause [*see illustration on next two pages*].

The earth is also encircled by belts of geological activity in the form of volcanoes, earthquakes and high heat flow, and observable motions in the form of folded rocks and the large displacements known as faults. In recent years the direction of displacements that are not observable on the surface has been deduced by the study of seismic waves arriving at various points on the earth's surface from earthquakes. It is now possible to tell the direction of slippage in the zones of rupture within the solid rocks of the earth's near-surface regions, so that the directions of the forces can be obtained.

If one looks at a map such as the one on the next two pages, one is immediately struck by the large scale and systematic distribution of these lines of geologic activity. Some of the systems are coherent over distances of several thousand kilometers. This immediately suggests the large-scale motion of material in the earth's interior. It does not, however, necessarily imply motions extending a similar distance into the interior. It is

apparently new, there are large blocks that have the same age as the cratons. It looks as if the earth's surface has been warped and folded around the ancient continental masses, catching up segments of the crust and intruding younger igneous rocks into the folds. In some places the ancient material has been altered beyond recognition, but elsewhere it has been left fairly undisturbed and its antiquity can be determined by radioactive-dating methods. These composite belts are termed zones of rejuvenation. When they are eroded down to sea level, all we see, as far as topography is concerned, is another part of the

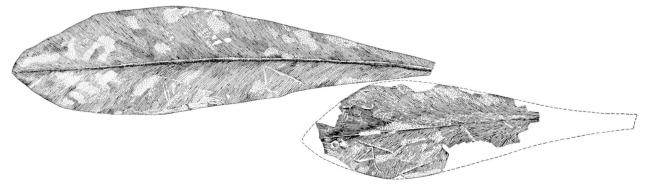

TYPICAL GONDWANA FLORA are *Glossopteris communis* (*leaf at left*) and *Gangamopteris cyclopteroides* (*right*), two species of fern that are identified in the Gondwana succession illustrated at the top of these two pages. The fossils from which these drawings were made were uncovered in the central part of Antarctica in 1961–1962 by William E. Long of Alaska Methodist University.

possible to have sheets of rigid material supporting stresses and fracturing over great distances if the underlying material is less rigid.

The topography of the ocean floors has been rapidly revealed in the past two decades by the sonic depth recorder. The principal systems of ridges and faults have been mapped in considerable detail by such oceanographers as Bruce C. Heezen and Maurice Ewing of Columbia University and H. W. Menard of the Scripps Institution of Oceanography. The layers of sediment on the sea floor have also been explored by such methods as setting explosive charges in the water and recording the echoes. It became a great puzzle how in the total span of the earth's history only a thin veneer of sediment had been laid down.

The deposition rate measured today would extend the process of sedimentation back to about Cretaceous times, or 100 to 200 million years, compared with a continental and oceanic history that goes back at least 3,000 million years. How could three-quarters of the earth's surface be wiped clean of sediment in the last 5 percent of terrestrial time? Furthermore, why were all the oceanic islands and submerged volcanoes so young? The new oceanographic investigations were presenting questions that were awesome to contemplate.

In the early 1960's Harry H. Hess of Princeton University and Robert S. Dietz of the U.S. Coast and Geodetic Survey independently proposed that the oceanic ridge and rift systems were created by rising currents of material

which then spread outward to form new ocean floors. On this basis the ocean floors would be rejuvenated, sweeping along with them the layer of sedimentary material. If such a mechanism were at work, no part of the ocean basins would be truly ancient. Although this radical hypothesis had much in its favor, it appeared farfetched to most.

Tracking the Shifting Poles

During this time a group of physicists and geophysicists were studying the directions of magnetism "frozen" into rocks in the hope of tracing the history of the earth's magnetic field. When an iron-bearing rock is formed, either by crystallization from a melt or by precipitation from an aqueous solution, it is

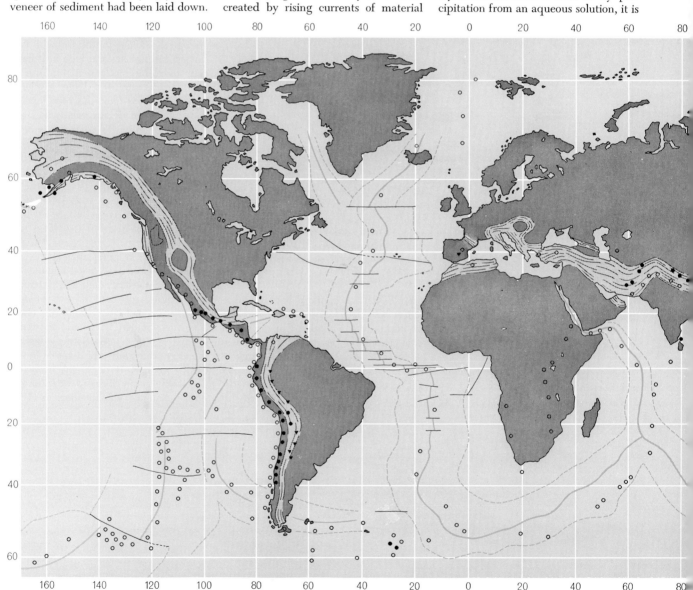

WORLDWIDE GEOLOGICAL PATTERNS provide evidence that the major land masses have been driven apart by a slow convection process that carries material upward from the mantle below the earth's crust. The dark-colored lines identify the crests of oceanic ridges that are now believed to coincide with upwelling regions. These ridges are crossed by large transcurrent fracture zones. The broken lines show the approximate limits of the oceanic rises. The light gray areas identify the worldwide pattern of recent mountain belts, island arcs, deep trenches, earthquakes and volcanism that apparently mark the downwelling of crustal material. The

slightly magnetized in the direction of the earth's magnetic field. Unless this magnetism is disturbed by reheating or physical distortion it is retained as a permanent record of the direction and polarity of the earth's magnetic field at the time the rock was formed. By measuring the magnetism in rocks of all ages from different continents, it has been possible to reconstruct the position of the magnetic pole in the past history of the earth. Great impetus was given to this study by P. M. S. Blackett and Runcorn, who with others soon found that the position of the pole followed a path going backward in time that was different for each continent [*see top illustration on next page*].

The interpretation of this effect was that the continents had moved with re-

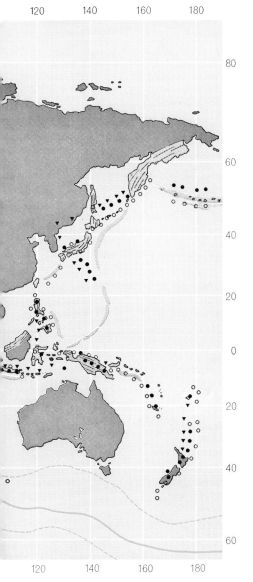

downwelling seems to coincide with the occurrence of deep earthquakes (*triangles*) and earthquakes of intermediate depth (*solid dots*). Upwelling zones seem to coincide only with shallow earthquakes (*open dots*).

spect to the present position of the magnetic pole, and that since the paths were different for each land mass, they had moved independently. Because it was unlikely that the magnetic pole had wandered very far from the axis of the earth's rotation, or that the axis of rotation had changed position with respect to the principal mass of the earth, it was concluded that the continents had moved over the surface of the earth. Moreover, since the shift in latitude of the southern continents was generally southward going backward in time, the motions were in accord with the older evidence pointing toward a Gondwanaland in the south-polar regions. In short, the magnetic evidence supported not only the notion of continental drift but also the general locations from which the continents had moved within the appropriate time span.

This was still not enough to sway the preponderance of American scientific opinion. Finally, at the annual meeting of the Geological Society of America in San Francisco in 1966, came the blows that broke the back of the opposition. Several papers put forward startling new evidence that related the concepts of ocean-floor spreading and continental drift, the cause of the oceanic-ridge and fault systems and the direction and time scale of the drift motions. In addition, the development of new mechanisms explaining displacement along faults brought into agreement some of the formerly contradictory seismic evidence.

In the study of rock magnetism it was observed that the earth's magnetic field not only had changed direction in the past but also had reversed frequently. In order to study how frequently and when the reversals occurred three workers in the U.S. Geological Survey—Allan Cox, G. Brent Dalrymple and Richard R. Doell—carefully measured the magnetism in samples of basaltic rocks that they dated by determining the amount of argon 40 in the rocks formed by the decay of radioactive potassium 40. They noted a distinct pattern of reversals over some 3.6 million years [see "Reversals of the Earth's Magnetic Field," by Allan Cox, G. Brent Dalrymple and Richard R. Doell; SCIENTIFIC AMERICAN, February, 1967]. Their finding was soon confirmed when Neil D. Opdyke and James D. Hays of Columbia University found the same pattern in going downward into older layers in oceanic sediments. It was thus established that the polarity of the magnetic field had universally reversed at certain fixed times in the past.

Meanwhile an odd pattern of magnetism in the rocks of the ocean floors had

been detected by Ronald G. Mason and Arthur D. Raff of the Scripps Institution of Oceanography. Using a shipborne magnetometer, they found that huge areas of the ocean floor were magnetized in a stripelike pattern. Putting together these patterns, the discovery of magnetic reversals and Hess's idea that the oceanic ridges and rifts were the site of rising and spreading material, F. J. Vine, now at Princeton, and D. H. Matthews of the University of Cambridge proposed that the hypothesis of the continuous creation of new ocean floors might be tested by examining the magnetic pattern on both sides of an oceanic ridge. The extraordinary discovery that the pattern was symmetrical with the ridge was demonstrated by Vine and Tuzo Wilson, who studied the two sides of a ridge next to Vancouver Island.

The history of the magnetic field going back into the past was laid out horizontally in the magnetism of the rocks of the sea floor going away from the ridge in both directions. It appeared that new hot material was rising from the rift in the center of the ridge and becoming magnetized in the direction of the earth's field as it cooled; it then moved outward, carrying with it the history of magnetic reversals. Since the dates of the reversals were known, the distance to each reversed formation gave the rate of spreading of the ocean floor [*see bottom illustration on next page*].

This important piece of work was quickly followed up by James R. Heirtzler, W. C. Pitman, G. O. Dickson and Xavier Le Pichon of Columbia, who have now shown that the ridges of the Pacific, Atlantic and Indian oceans all exhibit similar patterns. In fact, these workers have detected recognizable points in the history of magnetic reversals back about 80 million years, or in the Cretaceous period, and have drawn isochron lines, or lines of equal age, over huge strips of the ocean floors. Hence it is now possible to date the ocean floors and perceive the direction and rate of their lateral motion simply by conducting a magnetic survey over them. The implications for the study of drifting continents are immediately apparent.

These and other new findings do not unequivocally call for continental drift. It might be possible to have sea-floor spreading without drifting continents. Nonetheless, the directions and rates of motion for both sea-floor spreading and continental drift are entirely compatible. Above all, the principal objection to a hypothesis of continental movement has been removed.

Looking back, it is interesting to ob-

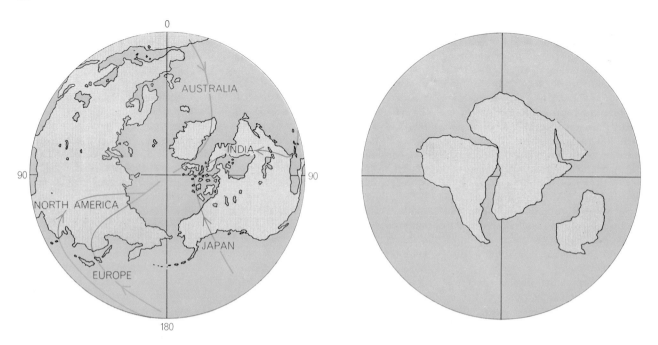

NORTH MAGNETIC POLE would appear to have wandered inexplicably during the past few hundred million years (*colored lines at left*), on the basis of "fossil" magnetism measured in rocks of various ages in various continents. The diagram is based on one by Allan Cox and Richard R. Doell of the U.S. Geological Survey. The pole could hardly have followed so many different tracks simultaneously; evidently it was the continents that wandered. K. M. Creer of the University of Newcastle upon Tyne found that the tracks could be brought together if South America, Africa and Australia were grouped in the late Paleozoic as shown at the right.

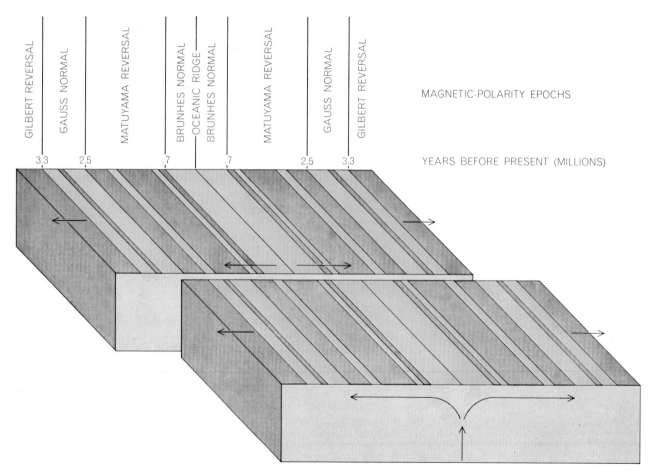

EVIDENCE FOR SEA-FLOOR SPREADING has been obtained by determining the polarity of fossil magnetism in rocks lying on both sides of oceanic ridges. In the diagram rocks of normal, or present-day, polarity are shown in color; rocks of reversed polarity are in gray. The displacement of the two blocks represents a transcurrent fracture zone. The symmetry suggests that the rocks welled up in a molten or semimolten state and gradually moved outward. The diagram is based on studies by a number of workers.

serve how each new piece of evidence presented in the past was met by counterevidence. Wegener's reconstruction, for example, was countered by numerous geologists who took exception to his detailed arguments. The arguments for the Permocarboniferous Gondwana glaciations were countered by Daniel I. Axelrod of the University of California at Los Angeles and others in this country. They contended that most species of fossil plant tend to be restricted to zones of latitude that hold for the continents in their present position, a fact that is hard to reconcile with the presumed pattern of glaciation. The idea that the great Gondwana land masses drifted in latitude has also been opposed by F. G. Stehli of Case Western Reserve University; his studies suggest that ancient fauna were most diverse at the Equator, and that the Equator defined in this way has not shifted.

Another Test of the Hypothesis

Any hypothesis must be tested on all points of observational fact. The balance of evidence must be strongly in its favor before it is even tentatively accepted, and it must always be able to meet the challenge of new observations and experiments. My own interest in the problem of continental drift was stimulated at a 1964 symposium in London sponsored by the Royal Society and arranged by Blackett, Bullard and Runcorn. At that time Bullard and his University of Cambridge associates J. E. Everett and A. G. Smith presented an elegant study of the geographic matching of continents on both sides of the North and South Atlantic. They had employed a computer to produce the best fit by the method of least squares. Instead of using shorelines, as had been done in earlier attempts, they followed the lead of S. W. Carey; he had chosen the central depth of the continental slope as representing the true edge of the continent.

The fit was remarkable [see illustration on page 56]. The average error was no greater than one degree over most of the boundary. My colleagues and I at the Massachusetts Institute of Technology now began to think of further testing the fit by comparing the sequence and age of rocks on opposite sides of the Atlantic.

Radioactive-dating techniques for determining the absolute age of rocks had reached a point where much could be learned about the age and history of both the ancient cratonic regions and the younger rejuvenated ones. For such purposes two techniques can be used in combination: the measurement of strontium 87 formed in the radioactive decay of rubidium 87 in a total sample of rock, and the measurement of argon 40 formed in the decay of potassium 40 in minerals separated from the rock. A collaborative effort was arranged between our geochronology laboratory and the University of São Paulo in Brazil (in particular with G. C. Melcher and U. Cordani of that institution). We also enlisted the aid of field geologists who had been working on the west coast of Africa (in Nigeria, the Ivory Coast, Liberia and Sierra Leone) and on the east coast of Brazil and Venezuela. The São Paulo group made the potassium-argon measurements of the Brazilian rock samples; we did the rubidium-strontium analyses on samples from all locations.

European geochronologists (notably M. Bonhomme of France and N. J. Snelling of Britain) had done pioneering work on the Precambrian geology of former French and British colonies and protectorates in West Africa. Of special interest to us at the start was the sharp boundary between the 2,000-million-year-old geological province in Ghana, the Ivory Coast and westward from these countries, and the 600-million-year-old province in Dahomey, Nigeria and east. This boundary heads in a southwesterly direction into the ocean near Accra in Ghana. If Brazil had been joined to Africa 600 million years ago, the boundary between the two provinces should enter South America close to the town of São Luís on the northeast coast of Brazil. Our first order of business was therefore to date the rocks from the vicinity of São Luís.

To our surprise and delight the ages fell into two groups: 2,000 million years on the west and 600 million years on the east of a boundary line that lay exactly where it had been predicted. Apparently a piece of the 2,000-million-year-old craton of West Africa had been left on the continent of South America.

In subsequent work on both sides we have found no incompatibilities in the age of many geological provinces on both sides of the South Atlantic [see illustration on next page]. Furthermore, the structural trends of the rocks also agree, at least where they are known. Minerals characteristic of individual belts of rocks are also found in juxtaposition on both sides; for example, belts of manganese, iron ore, gold and tin seem to follow a matching pattern where the coasts once joined.

Can such comparisons be made elsewhere? To some extent, yes. Unfortunately the rifting process by which a continent breaks up seems to be guided by zones of rejuvenation between cratons, as if these zones were also zones of weakness deep in the crust. It is necessary for the break to have transected the structure of the continent, cutting across age provinces, if one is to get a close refitting of the blocks. In the North Atlantic this is not the case, but the continental areas on both sides were simultaneously affected by an unmistakable oblique crossing of a Paleozoic belt of geological activity [see illustration on page 65]. Actually the belt covers the region of the Appalachian Mountains and the Maritime Provinces of North America, with an overlap along the coast of West Africa, and then splits into two principal belts: one extending through the British Isles and affecting the Atlantic coast of Scandinavia and Greenland and the other turning eastward into Europe. There is a superposition of at least four periods of renewed activity affecting the various parts of this complex. All four are represented on both sides of the North Atlantic, making this correlation extremely difficult to explain unless the continents were once together.

My colleagues H. W. Fairbairn and W. H. Pinson, Jr., and I, as well as other workers, have made age measurements in the northern Appalachians and Nova Scotia for many years, and we have found all four periods well represented in New England. The earliest period of activity (which Fairbairn has named Neponset) is dated about 550 million years ago; it is seen in some of the large rock bodies in eastern Massachusetts and Connecticut, in the Channel Islands off the northern coast of France, in Normandy, Scotland and Norway. The next-oldest period (the Taconic) was about 450 million years ago and is found on the western edge of New England and in parts of the British Isles. The next period, going back about 360 million years, is strongly represented in the entire span of the Appalachians and Nova Scotia (where it is called the Acadian) and in England and Norway (where it is called the Caledonian). Finally, about 250 million years ago, the activity seemed to move into southern Europe and North Africa, where it has been called the Hercynian. This activity, however, also extended into New England; much of southern Maine, eastern New Hampshire, Massachusetts and Connecticut show rocks of this age. Here the event is called the Appalachian.

Farther south the Lower Paleozoic section of the northwest coast of Africa (Senegal) appears to continue under the younger coastal sediments of Florida.

This African belt shows large rock units with ages equivalent to the Neponset, and also evidence of the younger events.

The Fitting of Antarctica

The recent extensive geological surveys in Antarctica have been highly rewarding in reconstructing Gondwanaland. Prior to the end of the Permian period the younger parts of western Antarctica were not yet formed. Only eastern Antarctica was present, including the great belts of folded rocks that form the Transantarctic Mountains. These consist of two geosynclines, or sediment-filled troughs: the inner Eopaleozoic and the outer Paleozoic [*see illustration on*

page 66]. The inner belt includes late Precambrian and early Cambrian sediments, which were folded and invaded by igneous rocks during late Cambrian or early Ordovician times (about 500 million years ago). Thus the inner belt is similar in age to the widespread event in the rest of Gondwanaland. It is marked by the Cambrian fossil Archaeocyatha, an organism that formed barrier reefs. These coral-like structures are found transecting sediments in bodies known as bioherms. The outer belt, farther within western Antarctica, is a geosyncline filled with Lower Paleozoic sediments. Like the northern Appalachians, it was deformed and invaded by igneous rocks in the middle and late Paleozoic.

Later it was covered with a quite representative Gondwana succession, with its glacial deposits, coal and diverse plants.

There seems to be a similar record of events in eastern Australia. The bioherms of the Cambrian Archaeocyatha are found in a belt extending northward from Adelaide and mark the edge of an early geosyncline filled with sediments including late Precambrian and Cambrian ones. Later in time, and farther to the east, great thicknesses of Silurian and Lower Devonian sediments accumulated in the Tasman trough. Compression and igneous intrusion occurred in this Tasman geosyncline mostly in the late Lower Devonian period to the middle Devonian (about 350 million years

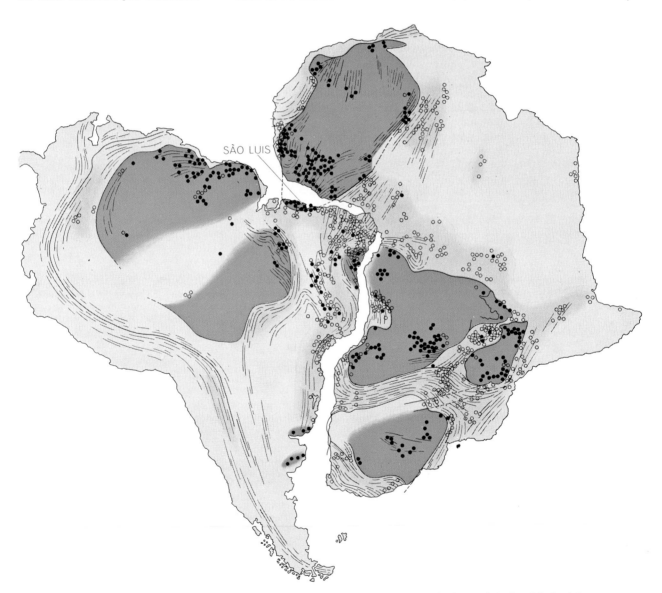

SÃO LUIS

TENTATIVE MATCHING of geological provinces of the same age shows how South America and Africa presumably fitted together some 200 million years ago. Dark-colored areas represent ancient continental blocks, called cratons, that are at least 2,000 million years old. Light-colored areas are younger zones of geological activity: mostly troughs filled with sediments and volcanic rocks that were folded, compressed and intruded by hot materials, forming granites and other rock bodies. Much of this activity was 450 million to 650 million years ago, but some of it goes back 1,100 million years. The dots show the sites of rocks dated by many laboratories, including the author's at the Massachusetts Institute of Technology. Solid dots denote rocks older than 2,000 million years; open dots denote younger rocks. The region near São Luís is part of an African craton left stranded on the coast of Brazil.

ago). The later cover of sediments includes a Gondwana succession similar to the one in Antarctica.

There is also strong evidence for a juncture between Australia and India, particularly in the Permian basins of sedimentation of the two continental blocks and in Gondwana sequences of coal and plants. Limestone beds containing the same Productid shells are found in the upper layers of the sequence on both sides. A correlation also exists between the banded iron ores of Yampi Sound in northwestern Australia and the similar ores of Singhbhum in India.

The illustration on page 66 is a reconstruction of Gondwanaland based on the evidence we have discussed so far. The three land masses—Antarctica, Australia and India—have been fitted together not at their present shorelines but where the depth of the surrounding ocean reaches 1,000 meters. As can be seen, the fit of the edges is good. The detailed fit of this assemblage into the southeastern part of Africa is still debated because most of the edges lack structures that cut across them. Nevertheless, I have included the edge of Africa in the map to show how it might possibly fit on the basis of limited age data from Antarctica.

This arrangement of land masses in the late Paleozoic is extremely tentative. It is now up to the geochronologists to test each juncture more closely for correlations in geologic age, and up to the field geologists to match structure and rock type. One particularly interesting fit may be forthcoming in a study of the boundaries of shallow and deep marine glacial deposits, and of the land tillites around what appears to be the start of an oceanic basin at the time Antarctica was breaking away. This attempt to establish the former position of Antarctica, which is being made by L. A. Frakes and John C. Crowell of the University of California at Los Angeles, may set in place the key piece in the puzzle. A detailed correlation of fossil plants in Antarctica with those of the adjacent land masses, which has been undertaken by Edna Plumstead of the University of Witwatersrand, is similarly limiting the possible position of the blocks.

The Age of the Atlantic

When did Gondwanaland begin to break up? One of the best pieces of evidence for the start of the opening of the South Atlantic is the age of offshore sediments along the west coast of Africa. Drilling through these sediments down to the ancient nonsedimentary rocks

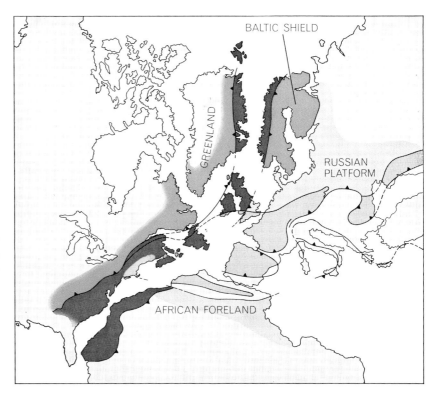

MATCHING OF NORTH ATLANTIC REGIONS is more difficult than in the South Atlantic. This tentative, pre-drift reconstruction of a portion of Laurasia depends on matching ancient belts of similar geological activity. The dark gray belt represents the formation of sediment-filled troughs and folded mountains in the early and middle Paleozoic (470 million to 350 million years ago). The medium gray belt was formed in the late Paleozoic (350 million to 200 million years ago). The latter belt overlapped the region of the former in the northern Appalachians and in southern Ireland and England, but diverged eastward in Europe. Four distinct and superimposed periods of geological activity occur on both sides of the present North Atlantic, providing strong evidence for a previous juncture.

shows that the layer of sediments is quite young: not older than the middle Mesozoic (about 160 million years ago). If the South Atlantic had been in existence for a major part of geologic time, the continent of Africa would unquestionably have developed a large shelf of sediments along the entire length of its western margin. The continental shelf would consist of sediments dating all the way back to the time of the ancient cratons. This is not the case. It looks as though the rift started from the northern edge of western Africa in the middle Triassic and slowly opened to the south until the final separation occurred in the Cretaceous. The east coast of Africa, on the other hand, apparently started to open earlier, in the Permian.

With the acceptance of sea-floor spreading and continental drift the global problems of geology are beginning to be solved. Although the train of thought on such matters is not universally accepted in detail, it is something like the following. Continental areas appear to have greater strength, to a depth of 100 kilometers or so, than ocean basins do, so that they tend to maintain themselves as buoyant masses that are not

destroyed by sinking motions. They can, however, be ruptured. Rising material pushes the surface apart; sinking material pulls the surface together and toward the region of sinking. Therefore if a sinking zone is established in an oceanic region, the continents will move toward the zone, and if a rising zone is established under a continent, the continent will split apart and the parts will move away from the zone. When the ocean floor moves toward a sinking zone in an oceanic region, it forms a deep trench bordered by volcanoes, chains of islands or elongated land masses such as the Philippines and Japan. When an ocean floor moves toward a continent, it appears to pass under the continental border, forming a great mountain chain. The mountain chain may be in part piled-up material that was already present and in part volcanic material that rose as the ocean swept its load of sediment, underlying volcanic rock and the continental shelf itself toward and under the edge of the continent. The process leads to a melting of underlying rock and to the intrusion of new volcanic material. The west coast of South America is a good example.

Another example is the thrust of India into Eurasia that formed the Himalayas. It has long been known that there was a large body of water between Africa and Eurasia and that a great thickness of sediments was deposited there at some time during the past 200 million years. This body is known as the Tethys Sea. It was located north of Arabia and extended from the former location of the

Atlas Mountains to east of the Himalayas. As I have mentioned, it appears that Gondwanaland not only broke up but also moved northward, with India and Africa pushing up into Eurasia. This motion apparently caused the buckling up of sediments in the Tethys Sea, giving rise to the mountain ranges that now form a contorted chain from the western Atlas range through the Medi-

terranean, the western Alps, the Caucasus and the Himalayas.

The way the present mountain systems of the earth fall along great circles suggests that the motions in the earth's interior have a large-scale coherence, of the order of the dimensions of the earth itself. The prevailing explanation stems from a new lead in seismology: a zone in the earth at a depth of 100 or 200 kilom-

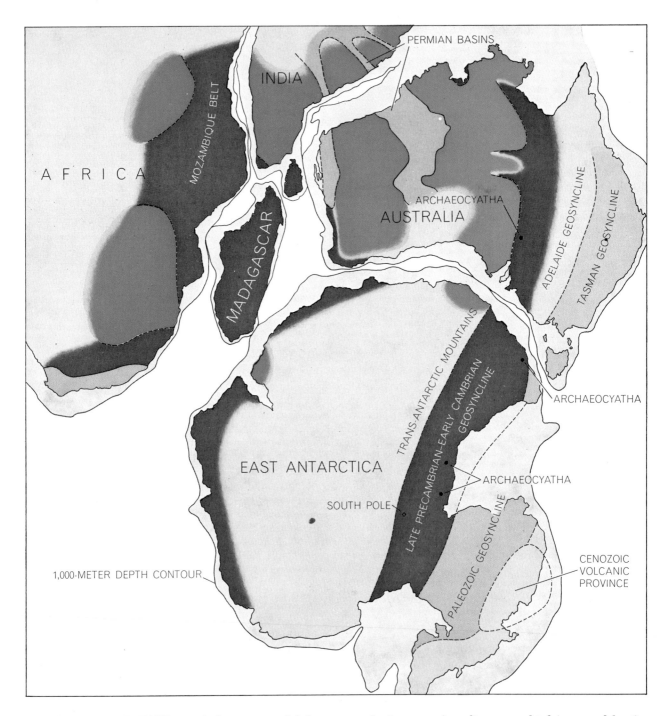

PART OF GONDWANALAND, tentatively reconstructed, brings together East Antarctica, Africa, Australia, Madagascar, India. The fit is at the 1,000-meter depth contour of the continental slope. Late Precambrian and Paleozoic geosynclines, or sediment-filled troughs, in eastern Australia are correlated in age and location with similar troughs along the Transantarctic Mountains. The deep Permian basins of northwest Australia match those of India. Glacial deposits, fauna and metal ores provide other correlations.

eters has been found to transmit seismic waves more slowly than the layers above and below it and to absorb seismic energy more strongly. This low-velocity zone is generally thought to consist of a material whose strength is reduced because a small amount of it is molten or because its temperature is approaching the melting point. The surface of the earth may therefore move around on this low-strength layer like the skin of an onion. It is believed the earth loses heat partly by conduction outward and partly by convection currents in the relatively thin layer above the weak zone. These currents, as they have been depicted by Walter M. Elsasser of Princeton and Egon Orowan of M.I.T., form rather flat convection cells.

A hypothesis that is currently popular is that the mechanism of spreading at the oceanic ridges involves the intrusion of hot material into ruptures near the surface. This material is the same as that in the low-velocity zone, lubricated by partly molten rock. A small proportion of the intruded material actually loses some of its melted fraction upward, giving rise to volcanoes and creating a thin layer (about five kilometers thick) of volcanic rock at the surface. The masses of intruded material cool as they move sideways from the central ridge, which is overlain by the thin layer of volcanic rock. This results in the observed distribution of seismic velocities at various depths, helps to explain why the flow of heat to the surface decreases with distance from the ridge and accounts for the pattern of magnetic reversals. At the sinking end of the convection cell this relatively rigid block of mantle material with its thin cover of basalt (plus a thin cover of new sediment) moves downward on an inclined plane.

It is clear where these concepts will lead. If folded mountain belts are the "bow waves" of continents plowing their way through ocean floors and ramming into other continents, we can use them to show us the relative directions of motion prior to the last great drift episode. If we look at the pre-drift Paleozoic mountain belts, such as the Appalachian belt of North America, the Hercynian of Europe and the Ural of Asia, we find that they are located *internally* in the great continental masses of Gondwanaland and Laurasia. This suggests that these pre-drift supercontinents had been formed by the inward motion of several separate blocks, which came together before they broke apart. Geologists have a new game of chess to play, using a spherical board and strange new rules.

8

SEA-FLOOR SPREADING

J. R. HEIRTZLER
December 1968

Comprehensive new theories that rationalize large numbers of observations and explain major aspects of the physical world are rare in any field of investigation. Such a synthesis may be within reach in geophysics. The past few years have seen the emergence of a new theory concerning systematic movements of the sea floor. It deals with vast and formerly unsuspected forces that churn the interior of the earth and account for the arrangement of ocean basins and land masses as we know them today. The theory is based on a variety of observations and hypotheses concerning the topography of the sea floor and the distribution of its sediments, the occurrence of faults and earthquakes, the internal structure of the earth, its magnetic field and periodic reversals of the field. It neatly supports the developing theory of continental drift. Together these theories have already been successful in explaining many surface features of the earth and providing information on internal earth processes. And it is possible that their full importance has yet to be appreciated—that they point toward a major synthesis relating the internal dynamics of the earth, its magnetic field and the dynamics of its orbital motions.

History of the Theory

The stage was set for the discovery of sea-floor spreading by the long debate over continental drift [see the articles "Continental Drift," by J. Tuzo Wilson, beginning on page 41, and "The Confirmation of Continental Drift," by Patrick M. Hurley, beginning on page 57]. Evidence from the shape, geological structure and paleontology of various continents and, within the past 20 years, studies of the "paleomagnetism" frozen into volcanic rocks had suggested that

the continents have drifted to their present locations from appreciably different positions in the course of millions of years. Even after the possibility of such drifting began to be recognized, it was not at all clear what forces could have caused great land masses to move over the surface of the globe.

By the late 1950's oceanographers had discovered that a continuous range of undersea mountains twists and branches through the world's oceans, that this ridge is usually found in the middle of the ocean and that earthquakes are associated with it. Marine geologists were aware too of the striking youth of the ocean floor: no bottom samples were ever found to be older than the Cretaceous period, which began some 135 million years ago. About 1960 Harry H. Hess of Princeton University proposed that the ocean floor might be in motion. He suggested a kind of convective movement that forced material from deep in the earth to well up along the axis of the mid-ocean ridges, to spread outward across the ocean floor and to disappear into trenches at the edges of continents. (The hypothesis seemed particularly attractive in the case of the Pacific Ocean, which is bordered by trenches, but it was less satisfactory for other oceans, which lack them.)

At about the same time Ronald G. Mason, Arthur D. Raff and Victor Vacquier of the Scripps Institution of Oceanography discovered that the ocean floor off the west coast of North America exhibited a remarkably regular striped pattern of variations in magnetic intensity [see "The Magnetism of the Ocean Floor," by Arthur D. Raff; SCIENTIFIC AMERICAN, October, 1961]. The pattern suggested great filamentary magnetic bodies, oriented north-south and offset at intervals along distinct lines running approximately at a right angle to the linea-

tions. No structural features that could explain such a pattern had ever been observed. The origin of these unique magnetic bodies remained a mystery for nearly five years. In 1963, after it had been noted that a distinct magnetic body could often be detected at the axis of a

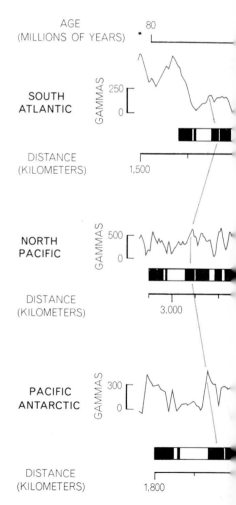

MAGNETIC ANOMALIES (*color*) recorded in all the world's oceans reveal the same succession of magnetic bodies (*black and white bands in strips*) parallel to the mid-ocean ridge. The bodies represent rock that

mid-ocean ridge, F. J. Vine and D. H. Matthews of the University of Cambridge proposed a convincing test of the hypothesis advanced by Hess. It was based on the discovery (which was just then being confirmed in detail) that the earth's magnetic field had reversed direction a number of times in past ages. They reasoned that if molten rock were pushed up along the axis of the mid-ocean ridge, it would become magnetized in the direction of the earth's prevailing magnetic field as it cooled. If the newly cooled material was subsequently pushed out away from the ridge, it would form strips of alternately "normal" and "reversed" magnetism, depending on the polarity of the earth's magnetic field when the rock solidified. A magnetometer at the surface of the ocean should detect these strips as positive or negative anomalies in the earth's smooth field.

Confirmation

As the Vine-Matthews proposal was being published, I was engaged, with colleagues from the Lamont Geological Observatory of Columbia University and the U.S. Naval Oceanographic Office, in a careful magnetic survey of the Reykjanes Ridge, a section of the Mid-Atlantic Ridge south of Iceland that was known to have large magnetic anomalies. We found that the anomalies were linear and symmetrically distributed parallel to the axis of the ridge. This strongly supported the idea of sea-floor spreading from the ridge and the formation of magnetic anomalies, just as Vine and Matthews had suggested. A little later Vine and J. Tuzo Wilson of the University of Toronto pointed out that the recent reversals of the field matched, one for one, part of the extensive pattern of magnetic lineations recorded just off the west coast of North America by Mason, Raff and Vacquier.

By 1965 it was clear to us and others that magnetism could be the key to reconstructing the history of the ocean floor and the movements of the continents. In only three years a great deal has been learned. Indeed, so many different workers have made significant contributions that it is impossible to name them all in a brief review or even always to know who was the first to make a new observation or propose a new model.

Pioneering efforts to measure the earth's magnetic field at sea had begun at Lamont about 20 years ago. Simple and precise instruments were designed to be towed behind ships and in time efficient techniques were devised for recording, storing and interpreting the data to delineate sea-floor structures hidden under layers of sediments. In 1965, when the new importance of magnetic anomalies became apparent, we had a large stock of data from all the oceans of the world and computer techniques with which to process the data. Examining the data in the light of the new hypotheses, we were able to recognize the same sequence of magnetic bodies extending away from the axis of the mid-ocean

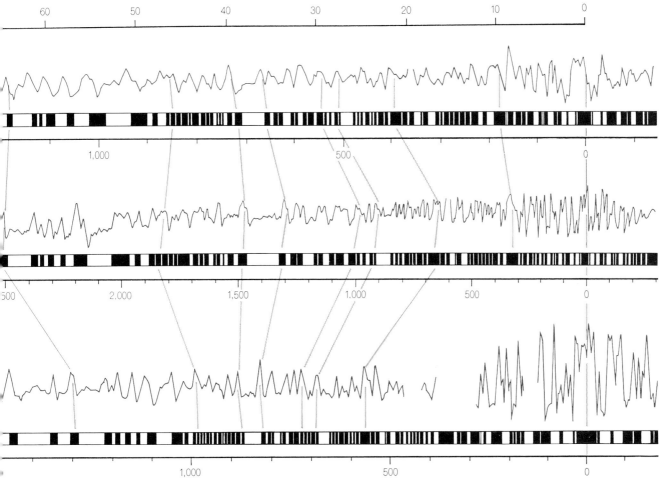

welled up along a "spreading axis" at the ridge during successive periods when the earth's magnetic field was "normal" as it is today or "reversed." The rock was magnetized in the ambient field and later forced out from the axis by subsequent flows. Here three magnetic traces are shown, from three oceans. The anomalies (in gammas, a measure of field intensity) and the magnetic bodies associated with each are spaced differently in each ocean because the spreading rates were different (the rate in the South Atlantic is believed to have been the most constant), but each ocean has the same sequence of 171 reversals extending back 76 million years.

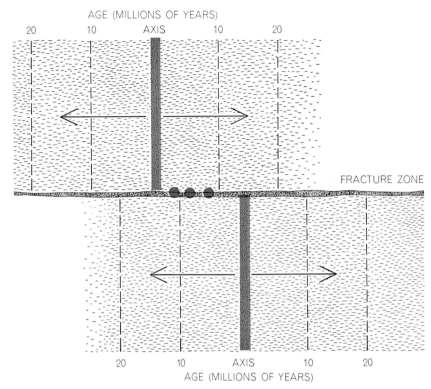

AGE (MILLIONS OF YEARS)

FRACTURE ZONE

AGE (MILLIONS OF YEARS)

MOLTEN ROCK wells up from the deep earth along a spreading axis, solidifies and is pushed out (*arrows*) by subsequent upwelling. The axis is offset by a fracture zone. Between two offset axes material on each side of the fracture zone moves in opposite directions and the friction between two blocks of the crust causes shallow earthquakes (*colored disks*).

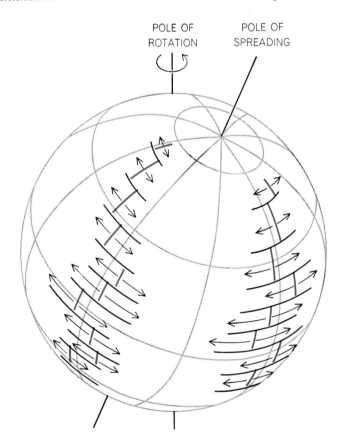

POLE OF ROTATION

POLE OF SPREADING

GEOMETRICAL RELATION between ridge axis and fracture zone becomes evident if one conceives of lines of latitude and longitude drawn about a "pole of spreading" rather than the pole of rotation. In each ocean the fracture zones are perpendicular to the spreading axis, and the rate of spreading (*arrows*) varies directly with distance from the equator.

ridge in the South Pacific, South Atlantic and Indian oceans.

Further examination has revealed a worldwide pattern of sea-floor spreading that makes sense of a wide variety of observations. It seems to explain the occurrence of many earthquakes. It establishes a detailed time scale for magnetic reversals and accounts for the direction and rate of continental drift. However, the precise geological events involved in the upwelling along the ridge and the downwelling at the edges of continents are still not understood in detail.

The Worldwide Pattern

At this point we know the main features of the spreading pattern in about half of the world's ocean areas, for the most part those that are adjacent to the better-explored sections of the mid-ocean ridge. The areas where the spreading pattern is unknown either are unexplored or simply resist explanation. For example, many oceans are not wide enough to show the very oldest magnetic bodies, and so not enough magnetic anomalies can be recorded to establish a rate of spreading. Some areas of the ocean floor show no magnetic anomalies. This may be because they are actually not part of the moving floor or because they were created in ancient geologic ages when the earth's field may not have been reversing. There may be places where the spreading is so slow that magnetic bodies crowd one another and confuse the magnetic-anomaly pattern. We have even found places where the axis of spreading is apparently under the continents or does not coincide with the axis of the mid-ocean-ridge system.

The mid-ocean ridge, the axis of the sea-floor spreading, does not meander smoothly from ocean to ocean; instead it is abruptly offset at many points. The offsets, or fracture zones, often extend to great distances on each side of the axis and are usually marked by some irregularity in the topography of the sea floor. Since material wells up at the axis and moves outward on both sides of a fracture zone, it is clear that there will be substantial friction along the zone between offset sections of the axis, where the material is moving in opposite directions [*see top illustration at left*]. Earthquakes, of course, are generated by just such rubbing and scraping between blocks of the earth's crust. Seismologists are now able to pinpoint the location and depth of an earthquake to within a few miles and also to identify the direction of initial motion of the crustal blocks in-

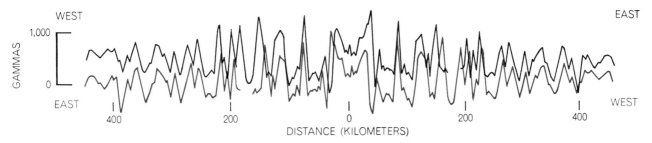

SYMMETRY of the magnetic record reflects the fact that similarly magnetized rock is pushed out on both sides of the axis. The symmetry is demonstrated by reversing a record covering about 1,000 kilometers (*color*) and superposing it on the original (*black*).

volved. Such measurements confirm the theory of sea-floor spreading: Most earthquakes along the ridge system occur in fracture zones between offset sections of the spreading axis—and the direction of first motion is just what one would predict on the basis of the sea-floor motion.

Although it does not appear so on a map with the usual Mercator projection, an axis of spreading and its associated fracture zones are at right angles. In the case of the Pacific and South Atlantic oceans the relation is particularly clear and includes even the rate of spreading. If one conceives of latitude and longitude lines drawn not about the earth's pole of rotation but about a "spreading pole," then the axis of spreading is parallel to the new lines of longitude and the fracture zones are perpendicular to the axis and parallel to the lines of latitude [*see bottom illustration on opposite page*]. Furthermore, the fastest spreading occurs along the equator of this new coordinate system and the rate decreases regularly with distance from the equator—as if giant splits in the floor of the oceans were taking place along the axes. The spreading poles vary for different oceans. For the Pacific and South Atlantic they appear to be very close to the earth's magnetic poles: near Greenland and in Antarctica south of Australia. Spreading in the Indian Ocean seems to be related to a different pair of poles, which can be located, with less certainty, in North Africa and in the Pacific north of New Zealand. Present information indicates that spreading in the North Atlantic occurs about a third set of poles, and fracture zones in older parts of the northeast Pacific floor seem to have been associated with yet another set of poles. There are indications that at least some of and perhaps all the spreading axes are themselves in motion, so that the location of their poles must be changing.

Some workers, notably W. Jason Morgan of Princeton, have attempted to generalize the motions of the earth's crust still further, suggesting that the crust of

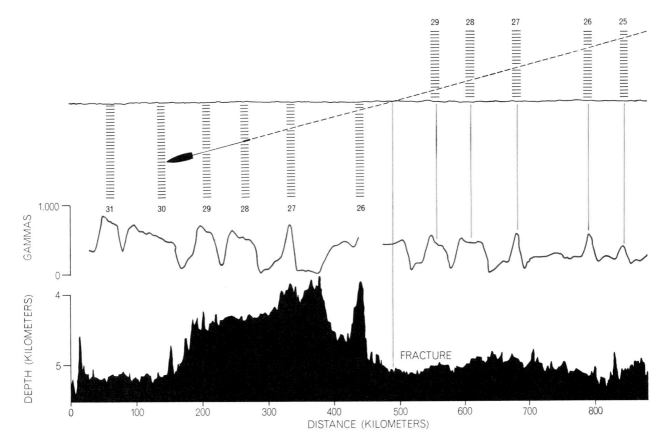

FRACTURE ZONE is identified by magnetic and bottom-profile data. A magnetometer was towed along a course (*broken line at top*) that twice crossed the same magnetic bodies (*Nos. 26–29*), offset along a fracture zone. The bodies caused similar magnetic anomalies, recognizable in the magnetic record (*middle*). Soundings revealed a prominent feature near the fracture zone (*bottom*).

the earth is divided into perhaps six rigid plates. These plates grow by the addition of new crust as new material wells up at the spreading axis; at their outer edges they override or are overridden by other plates. The concept of such strong and rigid plates is appealing to the seismolo-gist since, as will be explained later, deep earthquakes seem to be located at the proposed edges of many of them.

The Geomagnetic Time Scale

The magnetized bodies in the ocean floor have provided an amazingly complete history of magnetic-field reversals that now extends back about 76 million years, into the Cretaceous period. We began with a time scale that had been established by workers at Stanford University and the U.S. Geological Survey,

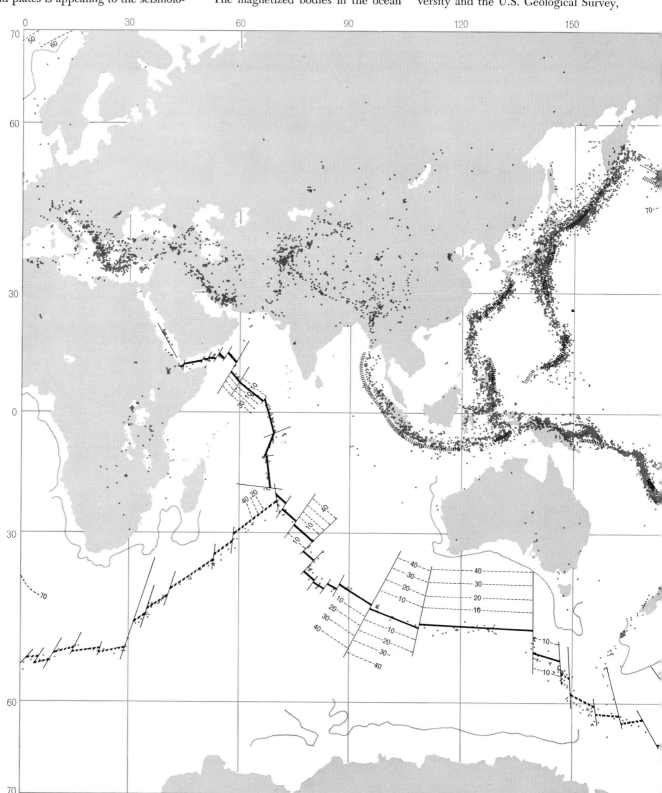

WORLDWIDE PATTERN of sea-floor spreading is evident when magnetic and seismic data are combined. Mid-ocean ridges (*heavy black lines*) are offset by transverse fracture zones (*thin lines*). On the basis of spreading rates determined from magnetic data the author and his colleagues established "isochrons" that give the age of the sea floor in millions of years (*broken thin lines*). The edges

who correlated the magnetism and radioactive-decay dates of rock samples going back some 3.5 million years [see "Reversals of the Earth's Magnetic Field," by Allan Cox, G. Brent Dalrymple and Richard R. Doell; SCIENTIFIC AMERICAN, February, 1967]. By comparing the ages they had assigned to specific reversals of the geomagnetic field with the distance of the corresponding anomalies from the ridge axis, we were able to extend the observations of Vine and Wilson over much of the ocean floor and thus to determine the rate at which the floor had spread in the various oceans.

The rates were found to be different for different sections of the ridge but seemed to have been remarkably steady over the ages in many areas. They vary from about half an inch to a little more

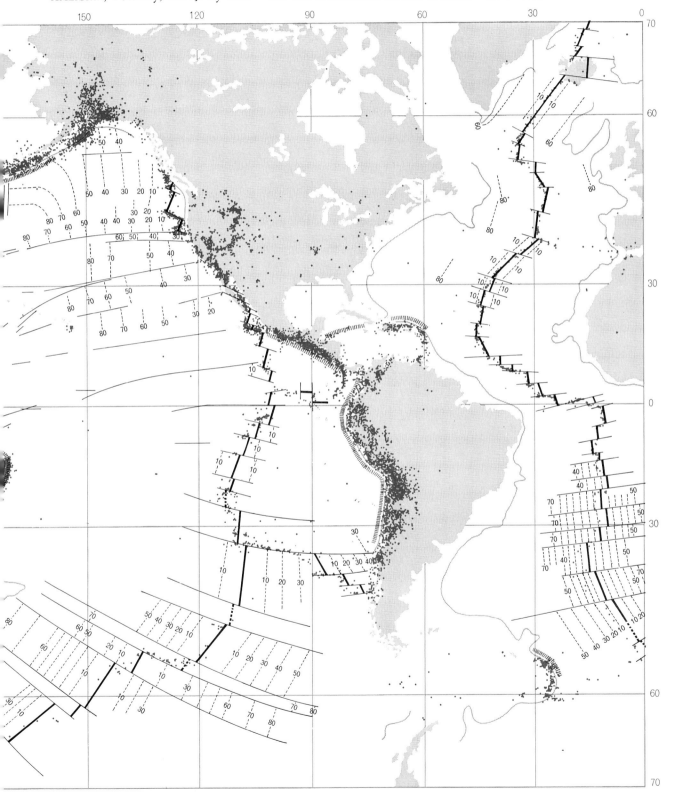

of many continental masses (*gray lines*) are rimmed by deep ocean trenches (*hatching*). When the epicenters of all earthquakes recorded from 1957 to 1967 (plotted by Muawi Barazangi and James Dorman from U.S. Coast and Geodetic Survey data) are superposed (*colored dots*), the vast majority of them fall along mid-ocean ridges or along the trenches, where the moving sea floor turns down.

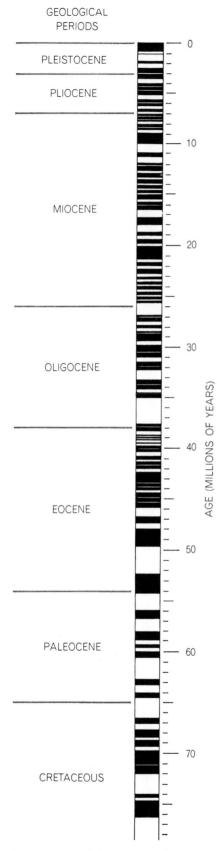

GEOLOGICAL
PERIODS

PLEISTOCENE

PLIOCENE

MIOCENE

OLIGOCENE

EOCENE

PALEOCENE

CRETACEOUS

AGE (MILLIONS OF YEARS)

0

10

20

30

40

50

60

70

CHRONOLOGY of geomagnetic-field reversals was worked out by extrapolation, beginning with the dates that had been established for recent reversals and assuming a constant rate for spreading within each ocean. The dates were confirmed by geological data.

than two inches per year away from the axis of spreading. Although these rates of movement are small by everyday standards, they are large on a geological time scale. Their magnitude came as a surprise to geophysicists even though comparable rates of slippage had been observed along the San Andreas fault, an exposed fracture zone in southern California.

In several areas we saw that there was no obvious hiatus in spreading. We therefore felt free to assume a constant spreading rate and on that basis to assign dates to geomagnetic field reversals extending far beyond the 3.5-million-year scale. This extrapolation of spreading rates may not seem justified at first, but it is supported by consistency from ocean to ocean and by agreement with other geophysical evidence, such as radioactive and paleontological dating of rock samples. Within the probable errors of the methods no discrepancies have been found. We have now identified 171 reversals of the earth's magnetic field over the 76 million years and believe that the dates assigned to each are quite accurate. (Of course, if spreading stopped abruptly all over the world for a certain time, our time scale before the spreading stopped would be too young by that number of years; this is improbable but cannot be entirely ruled out.) On the whole the evidence for the correctness of the time scale is so strong that it can now be used in turn to study variations in spreading rates over the ages in certain disturbed areas.

The average "normal" interval (when the magnetic field was oriented as it is today) works out to 420,000 years and the average "reversed" interval to 480,-000 years. The closeness of the two numbers means that the earth is just about as likely to be found in one state of magnetic polarity as in the other. The present era of normal polarity has lasted 700,000 years. Are we due for a change? Only 15 percent of the normal intervals have been longer than that, although some were apparently as long as three million years. The shortest intervals seem to have been less than 50,000 years, but brief instants of geologic time are difficult to confirm by absolute dating methods. This suggests one of the drawbacks of the magnetic scale: simply because it is so detailed, it is unlikely soon to be proved wrong; by the same token, however, it is difficult to use to date a piece of magnetized earth material. Just as one may need a "finder" telescope to orient a high-powered telescope, so one must start by knowing the approximate age of

materials in order to utilize the geomagnetic time scale.

The geomagnetism I have been discussing is frozen into igneous rock of the basaltic type that wells up from deep within the earth. Over much of the ocean floor, of course, this bedrock is not exposed but is covered by varying thicknesses of accumulated sediments. These sediments can also be magnetized, since the tiny particles of which they are composed can become oriented in the earth's field as they settle slowly to the ocean floor. (They are magnetized only one ten-thousandth as strongly as the basalt, however, so that even a thick layer of sediment does not interfere with measurements of anomalies caused by the underlying basalt.) By dropping a hollow cylindrical pipe into the bottom mud one can bring to the surface long "cores" of successive layers of sediment, each constituting an undisturbed record of magnetic-field reversals. Workers at Scripps and Lamont have recently developed sensitive techniques for measuring these weakly magnetized specimens, and the record of geomagnetic reversals has been verified back about 10 million years. The ability to correlate sedimentary layers of the same age in different oceans has proved of immense value to marine geologists, who often lack good paleontological or other indicators of geological horizons. In making detailed comparisons of magnetic reversals and the populations of microscopic animal fossils in such cores, investigators have noted a striking correlation between reversals and major changes in microfaunal species. It has been proposed that such changes are the result of mutations caused by increased exposure to cosmic rays if the earth's protective magnetic field is largely attenuated during a reversal. The evidence for this is not strong, however, and an alternative explanation can be put forward, as I shall suggest later in this article.

Footprints of Continents

The indicated movements of the ocean floor are of the right size and in the right directions to account for continental drift. Topographic and geological evidence had pointed to the probable existence some 200 or 300 million years ago of a single large land mass in the Southern Hemisphere (named Gondwanaland for a key geological province in India) that included Africa, India, South America, Australia, New Zealand and Antarctica, and another mass in the Northern Hemisphere called Laurasia. The positions of the present continents within these land masses were not clear, nor was it possible to trace in detail the sequence of events in their breakup.

The magnetic lineations in the ocean floor serve as "footprints" of the continents, marking their consecutive positions before they reached their present locations. We found that the slow but steady and prolonged rates of motion were sufficient to separate South America from Africa—thus creating the entire South Atlantic Ocean—in about 200 million years and to separate Australia from Antarctica in about 40 million years. As more of the sea floor was dated we could establish more exactly just when the various continents separated and how they moved [see illustration at right]. It is now possible to reconstruct the original positions of the continents and the shallow continental shelves, so that land geologists can go into the field and check the continuity of ancient geological structures that were torn apart when the separations occurred.

Although it is possible to tell when and how the continents pulled away from one another, it is not always entirely clear whether or not—and for how long—a continent may have stood still. The impression is that both Africa and Antarctica have remained fixed with respect to the rotational axis for the last 100 million years, or since Africa split off from the remainder of Gondwanaland. If this is true, then the fact that the spreading axis between South America and Africa remains about midway between them indicates that the axis itself must be moving too.

Source and Sink

Until the theory of sea-floor spreading was advanced the only known sources of deep-earth material at the surface were volcanic eruptions. It now appears, however, that most currently active volcanoes are not at the sites of upwelling along the mid-ocean ridge but rather in areas where the moving sea floor turns down under the continents. Recent volcanic eruptions in the Philippines, Mexico and Guatemala are examples, as is the continuing intermittent activity of volcanoes in western North America, Alaska and Japan. The upwelling that initiates sea-floor spreading must therefore represent some unfamiliar geophysical phenomenon, and investigators have concentrated much attention on the axis of the mid-ocean ridge where it occurs. The axis has several unusual properties: a large heat flux, a concentration of shallow earthquakes, unusual seismic-wave velocities, a lack of sediment and eroded rocks, and the presence of a prominent magnetic anomaly. While most of these unusual conditions extend over a distance of from tens to hundreds of miles on both sides of the crest of the ridge, the magnetic anomaly is sharply localized; by studying the symmetry of the magnetic pattern it is possible to locate the spreading axis to within a few miles. The terrain of the ridge is rugged and it is not possible to associate the spreading axis with any single topographic feature,

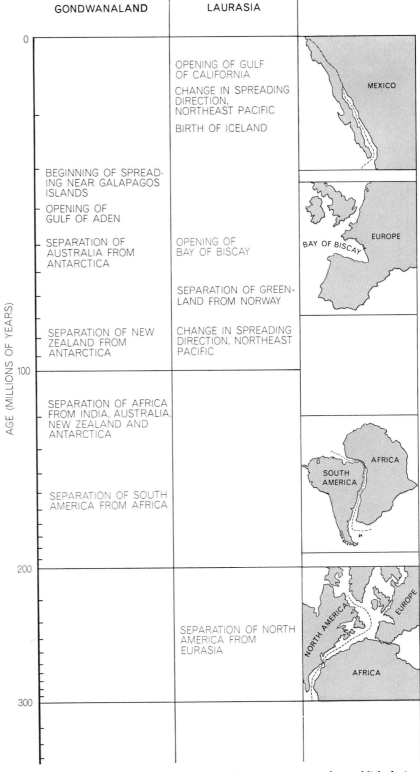

TIME SCALE for continental movements and other changes can now be established, since isochrons show the age of the ocean floor and the direction of spreading at any time.

although one can say that when the ridge contains a median "rift valley," as many do, the axis lies within it.

The magnetized plate is not very thick. Analyses of the observed magnetic anomalies suggest that the thickness is from a half-mile to a few miles; studies of the transmission of seismic waves and the distribution of heat flux, on the other hand, indicate a thickness of a few tens of miles or perhaps 100 miles for the moving plate. The precise linearity of the magnetized bodies shows that the upwelling material was not tumbled about after it became cooled and magnetized; in this it is different from the usual folded rocks seen on land or the irregular flows that surround a volcanic eruption. Attempts to locate the axis with a magnetometer submerged near the sea floor show that there are a number of very magnetic bodies where the axis would be expected.

The evidence suggests to most investigators that the upwelling mechanism is an injection of molten deep-earth material by linear intrusions called dikes. Such bodies have a high probability of being injected along the line of the spreading axis. Each injection may be quite localized rather than greatly elongated along the axis. The new material is hot enough to reheat adjacent rock so that both the new material and the slightly older material nearby is magnetized in the direction of the ambient field before being quenched by the cold seawater. This explanation does not indicate anything precise about the thickness of the moving layer; it does account for the lack of tumbling of magnetized blocks, since a dike would tend to push the older material horizontally away from the ridge.

The deep oceanic trenches found around the periphery of the Pacific are thought to be places where surface material returns to the deep earth; so, very likely, are smaller trenches in the North Atlantic and South Atlantic and the Indian Ocean. In many parts of the world where plates of moving sea floor have been identified the outer edge of the plate has not been located, and so we probably do not know where all the sinks are. In most places the spreading sea floor seems to be turning down under the continents, but in some places it seems to be pushing continents ahead of it or even tearing continents apart; we know of no place where the spreading floor is overriding a continent.

If the epicenters of deep earthquakes are mapped, they are almost all found to be on a plane that starts at the floor of a trench and makes an angle of about 45 degrees with the horizontal [see top illustration on page 77]. The slippage of earth material parallel to the plane and extending hundreds of miles below the surface is the cause of the earthquakes. These earthquake planes almost certainly define regions of downwelling; studies of first motion in such areas show that the sea floor moves down with respect to the adjacent continent. A thin crustal layer with a characteristic speed of seismic-wave propagation has been shown to underlie the ocean sediments and turn down at 45 degrees to great depths.

The magnetic evidence for what happens at the trenches is ambiguous, however. The sea-floor magnetic pattern is altered suddenly. Measurements at the Aleutian Trench, for example, show a

ANCIENT POSITIONS of the continents can be plotted on the basis of the time scale. This map shows the relation of the Americas to Eurasia and Africa 70 million years ago. The continents fitted together generally along the edges of continental masses (*light grey*) rather than of present dry-land areas (*dark gray*). Broken lines trace directions of movement.

sharp discontinuity [*see bottom illustration on this page*]. There is no sign of a magnetic body that should (on the basis of measurements made elsewhere) be located about three kilometers below the trench floor. Its absence might be explained by heating or mechanical deformation, but neither sufficiently high temperatures nor enough seismic activity to cause deformation are indicated so close to the trench floor. Another problem at the Aleutian Trench is that the magnetic bodies seem to be in the wrong sequence. If the trenches are sinks, the older bodies should be in the trench and the younger ones to seaward; the opposite seems to be the case! Moreover, south of the Alaska Peninsula bodies oriented essentially north-south turn and run east-west, parallel to the coast. This complexity is difficult to explain if trenches are sinks. It is one of the most awkward sticking points in the theory.

Speculations

The feeling among many geophysicists that we are on the brink of even more comprehensive theories about the earth stems from the striking geographic or temporal coincidence of certain geophysical phenomena, many having to do with the dynamics of our planet. Existing theories suggest no clear causal relation among these phenomena, and so one hopes for some higher-order synthesis that will establish such a relation.

What are some of these coincidences? The present pole of spreading (for several of the oceans, at least) is near the geomagnetic axis; in Cretaceous times the spreading pole for the North Pacific was near the Cretaceous geomagnetic pole. There have often been significant changes in the microfaunal population of the sea at magnetic reversals. A major meteorite (tektite) fall occurred just at the time of the last reversal [see "Tektites and Geomagnetic Reversals," by Billy P. Glass and Bruce C. Heezen; SCIENTIFIC AMERICAN, July, 1967]. Some authors have speculated recently on a relation between mountain-building activity and magnetic reversals; others see a relation between changes in sea-floor spreading and mountain-building. Mechanisms have been suggested whereby the earth's magnetic field could be generated by convective motion caused in turn by irregularities in the earth's orbit. There has been a revival of a 30-year-old theory that the glacial ages were caused by changes in the tilt of the earth's axis. Finally, there is clear evidence that large earthquakes occur

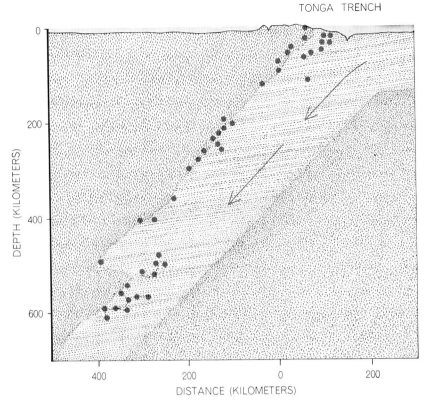

MOVING SEA FLOOR turns down at the trenches rimming the Pacific Ocean. Deep earthquake epicenters line the downward path where floor material moves under land masses. Symbols (*color*) show epicenters recorded at the Tonga Trench in the South Pacific.

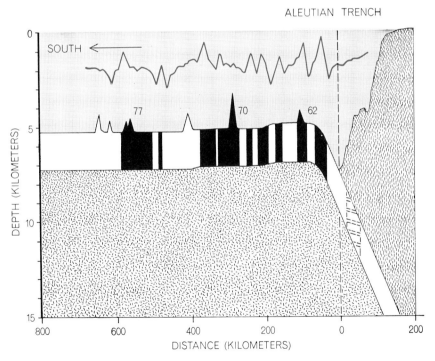

ALEUTIAN TRENCH data fail to fit the model, however. The magnetic record (*color*) shows the expected sequence of normal and reversed bodies (*black and white bands*) approaching the trench, but there is no evidence of the next body, which should appear about three kilometers below the trench (*hatched band*). Moreover, the magnetic bodies seem to be younger (*77 to 62 million years*) as one approaches the trench rather than older.

at about the same time as certain changes in the earth's rotational motion.

This article is not the place for a full evaluation of all these developments. It is interesting to note, however, that a common thread running through these and many other proposals at the frontier of geophysical research is the role played by displacements of the earth's axis of rotation. It seems that rather minor variations can affect to a surprising extent both the climate at the surface of the earth and forces and stresses within the earth.

The intimate relation between the spreading pole and the magnetic pole suggests that the convective motion within the earth and the earth's field may have a common cause. It could be due to the proposed effect of orbital irregularities or to some dynamo action within the earth. Since the direction of convec-tion does not reverse when the field reverses, it is clear that the convective motion does not simply generate the field, nor is it likely that the reversing field could "pump" convection cells. Whatever the driving force is for these two phenomena, it would seem to be related to the motion of the earth. Recently it has been shown that earthquakes of a magnitude of 7.5 or greater on the Richter scale either cause or are caused by changes in the earth's "wobble," a small circular motion described by the north pole of rotation [see "Science and the Citizen," SCIENTIFIC AMERICAN, Nov., 1968]. Whatever the mechanism of these changes, it is not hard to believe that similar changes in the earth's axial motion in times past could have caused major earthquake and mountain-building activity and could even have caused the magnetic field to flip.

To summarize: Every few months there are changes in the earth's rotational motion that affect sea-floor spreading and cause the earthquakes associated with it. If such a change is large enough, it may even reverse the earth's magnetic field. (Both the presence of a geomagnetic field and the spreading of the sea floor seem to be due to the mere fact that the earth is rotating; it is only the changes in motion that are associated with certain large earthquakes and reversals of the field.) Changes in the microfaunal population of the sea are related to changes in the climate, which are related in turn to variations in the earth's motions.

Such speculation cannot yet be confirmed but neither can it be firmly denied; indeed, it is no more outlandish than theories of sea-floor spreading seemed to be a few years ago.

9

THE DEEP-OCEAN FLOOR

H. W. MENARD
September 1969

Oceanic geology is in the midst of a revolution. All the data gathered over the past 30 years—the soundings of the deep ocean, the samples and photographs of the bottom, the measurements of heat flow and magnetism—are being reinterpreted according to the concept of continental drift and two new concepts: sea-floor spreading and plate tectonics (the notion that the earth's crust consists of plates that are created at one edge and destroyed at the other). Discoveries are made and interpretations developed so often that the scientific literature cannot keep up with them; they are reported by preprint and wandering minstrel. At such a time any broad synthesis is likely to be short-lived, yet so many diverse observations can now be fitted into a coherent picture that it seems worthwhile to present it.

Before continental drift, sea-floor spreading and plate tectonics captured the imagination of geologists, most of them conceived the earth's crust as being a fairly stable layer enveloping the earth's fluid mantle and core. The only kind of motion normally perceived in this picture was isostasy: the tendency of crustal blocks to float on a plastic mantle. The horizontal displacement of any geologic feature by as much as 100 kilometers was considered startling. This view is no longer consistent with the geological evidence. Instead each new discovery seems to favor sea-floor spreading, continental drift and plate tectonics. These concepts are described elsewhere in this book [see "The Origin of the Oceans," by Sir Edward Bullard, beginning on page 88]. Here I shall recapitulate them briefly to show how they are related to the actual features of the deep-ocean floor.

According to plate tectonics the earth's crust is divided into huge segments afloat on the mantle. When such a plate is in motion on the sphere of the earth, it describes a circle around a point termed the pole of rotation (not to be confused with the entire earth's pole of rotation). This motion has profound geological effects. When two plates move apart, a fissure called a spreading center opens between them. Through this fissure rises the hot, plastic material of the mantle, which solidifies and joins the trailing edge of each plate. Meanwhile the edge of the plate farthest from the spreading center—the leading edge—pushes against another plate. Where that happens, the leading edge may be deflected downward so that it sinks into a region of soft material called the asthenosphere, 100 kilometers or more below the surface. This process destroys the plate material at the same rate at which it is being created at its trailing edge. Many of the fissures where plate material is being created are in the middle of the ocean floor, which therefore spreads continuously from a median line. Where the plates float apart, the continents, which are embedded in them, also drift away from one another.

The most obvious consequence of this process on the ocean floor is the symmetrical seascape on each side of a spreading center. As two crustal plates move apart (at a rate of one to 10 centimeters per year), the basaltic material that wells up through the spreading center between them splits down the middle. The upwelling in some way produces a ridge, flanked on each side by deep ocean basins and capped by long hills and mountains that run parallel to the crest. The flow of heat from the earth's interior is generally high along the crest because dikes of molten rock have been injected at the spreading center. A spreading center may also open under a continent. If it does, it produces a linear deep such as the Red Sea or the Gulf of California. If it continues to spread, or if the spreading center opens in an existing ocean basin, the same symmetrical seascape is ultimately formed.

This symmetry extends to less tangible features of the ocean floor such as the magnetic patterns in the basalt of the slopes on either side of the mid-ocean ridge. As the plastic material reaches the surface and hardens, it "freezes" into it the direction of the earth's magnetic field. The earth's magnetic field reverses from time to time, and as each band of new material moves outward across the ocean floor it retains a magnetic pattern shared by a corresponding band on the other side of the ridge. The result is a matching set of parallel bands on both sides. These patterns provide evidence of symmetrical flows and make it possible to date them, since they correspond to similar patterns on land that have been reliably dated by other means.

The steepness of the mid-ocean ridge is determined by a balance between the rate at which material moves outward from the spreading center and the rate at which it sinks as it ages after solidifying. The rate of sinking remains fairly constant throughout the ocean basin, and it seems to depend on the age of the

PILLOW LAVA (as pictured on the following page) assumes its rounded shape because it cools rapidly in ocean water. This flow lies on the western slope of the mid-ocean ridge in the South Atlantic at a depth of 2,650 meters. Such flows erupt from the many volcanic vents and fissures that are created as the ocean floor spreads out from the mid-ocean ridges in the form of vast crustal plates. The photograph was made under the direction of Maurice Ewing of the Lamont-Doherty Geological Observatory.

crust. It can be calculated if the age of the oceanic crust (as indicated by the magnetic patterns) is divided into the depth at which a particular section lies. Such calculations show that the crust sinks about nine centimeters per 1,000 years for the first 10 million years after it forms, 3.3 centimeters per 1,000 years for the next 30 million years, and two centimeters per 1,000 years thereafter. Not all the crust sinks: on the southern Mid-Atlantic Ridge the sea floor has remained at the same level for as long as 20 million years.

The rate at which the sea floor spreads varies from one to 10 centimeters per year. Therefore fast spreading builds broad elevations and gentle slopes such as those of the East Pacific Rise. The steep, concave flanks of the Mid-Atlantic Ridge, on the other hand, were formed by slow spreading.

Whether the slopes are steep or gentle, the trailing edge of the plate at the spreading center is about three kilometers higher than the leading edge on the other side of the plate. The reason for this difference in elevation is not known. Heating causes some elevation and cooling some sinking, but the total relief appears much too great to be attributed to thermal expansion. Cooling might account for the relatively rapid sinking observed during the first 10 million years, but continued sinking remains a puzzle.

A decade ago scanty information suggested that the mid-ocean ridge in both the Atlantic and Pacific was continuous, with a few branches. More complete surveys have revealed that crustal plates have ragged edges. Instead of extending unbroken for thousands of kilometers, a mid-ocean ridge at the trailing edges of two crustal plates forms a zigzag line consisting of many short segments connected by fracture zones to other ridges, trenches, young mountain ranges or crustal sinks. The fracture zones connecting the ridge segments are associated with what are called "transform" faults. They provide important clues to the history of a plate. Because they form some of the edges of the plate, they delineate the circle around its pole of rotation, thereby indicating the direction in which it has been moving.

From what has been said so far it might appear that the spreading centers are fixed and stationary. The constantly repeated splitting of the new crust at the spreading center produces symmetrical continental margins, symmetrical magnetic patterns on the ocean floor, symmetrical ridge flanks and even

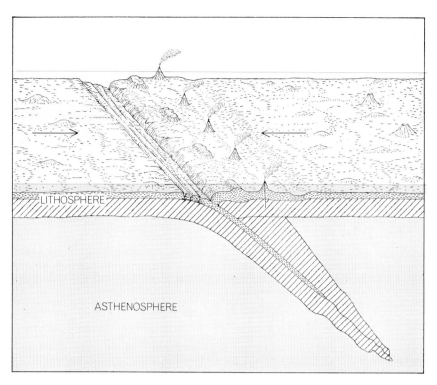

TRENCH IS CREATED where the leading edge of a plate that emerges from a fast spreading center collides with another plate. Because the combined speed of the two is more than six centimeters per year neither can absorb the impact by buckling. Instead one crustal plate (*in lithosphere*) plunges under the other to be destroyed in the asthenosphere, a hot, weak layer below. The impact produces volcanoes, islands and a deep, such as the Tonga Trench. Beside a trench are cracks that are produced by bending of the crust.

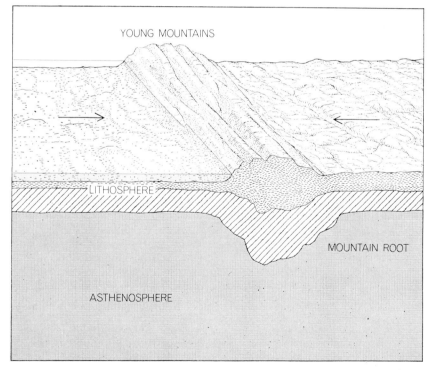

MOUNTAIN RANGE IS FORMED when the leading edges of two plates come together at less than six centimeters per year. Instead of colliding catastrophically, so that one plate slides under the other, both plates buckle, raising a young mountain range between them. The range consists of crustal material that folds upward under the compression exerted by the two plates (and also downward, forming the root of the mountain). Such ranges can be identified because they contain cherts and other material typical of the ocean bottom.

symmetrical mountain ranges. More often than not, however, it has been found that the spreading center itself moves. Oddly enough such movement gives rise to the same symmetrical geology. All that is required in order to maintain the symmetry is that the spreading center move at exactly half the rate at which the plates are separating. If it moved faster or slower, the symmetry of the magnetic patterns would be destroyed.

Imagine, for instance, that the plate to the east of a spreading center remains stationary as the plate to the west moves. Since the material welling up through the fissure splits down the middle, half of it adheres to the stationary plate and the other half adheres to the moving plate. The next flow of material to well up through the split thus appears half

the width of the spreading center away from the stationary plate. The flow after that appears a whole width of the spreading center away from the stationary plate, and so on. In effect the spreading center is migrating away from the stationary plate and following the moving one. If the speed of the spreading center exceeded half the speed of the migrating plate, however, a kind of geological Doppler effect would set in: the bands of the magnetic pattern would be condensed in the direction of the moving plate, and they would be stretched out in the direction of the stationary plate [see illustration below].

It might seem unlikely that the spreading center would maintain its even rate of speed and remain exactly between the two plates. W. Jason Morgan of Princeton University observes, how-

ever, that there is no impediment to such motion, provided only that the crust splits where it is weakest (which is where it split before, at the point where the hot dike was originally injected). As a result the spreading center is always exactly between two crustal plates whether it moves or not.

Moving spreading centers account for some of the major features of the ocean floor. The Chile Rise off the coast of South America and the East Pacific Rise are adjacent spreading centers. Since there is no crustal sink between them, and new plate is constantly being added on the inside edge of each rise, at least one of the centers must be moving, otherwise the basin between them might fold and thrust upward into a mountain range or downward into a trench. Similarly, the Carlsberg Ridge in the Indian Ocean and the Mid-Atlantic Ridge are not separated by a crustal sink and hence one of them must be moving.

A moving center may have created the ancient Darwin Rise on the western edge of the Pacific basin and also the modern East Pacific Rise. As in the case of the Carlsberg Ridge and the Mid-Atlantic Ridge, the existence of two vast spreading centers on opposite sides of the ocean with no intervening crustal sink has puzzled geologists. If such centers can move, however, it is possible that the spreading center in the western Pacific merely migrated all the way across the basin, leaving behind the ridges of the Darwin Rise. In this way one rise could simply have become the other. Many other examples exist, and Manik Talwani of the Lamont-Doherty Geological Observatory proposes that all spreading centers move.

As a plate forms at a spreading center it consists of two layers of material, an upper "volcanic" layer and a lower "oceanic" one. Lava and feeder dikes from the mantle form the volcanic layer; its rocks are oceanic tholeiite (or a metamorphosed equivalent), which is rich in aluminum and poor in potassium. The oceanic layer is also some form of mantle material, but its precise composition, density and condition are not known. Farther down the slope of the ridge the plate acquires a third layer consisting of sediment.

The sediment comes from the continents and sifts down on all parts of the basin, accumulating to a considerable depth. It is mixed with a residue of the hard parts of microorganisms that is called calcareous ooze. Below a certain depth (which varies among regions) this

SPREADING CENTER MOVES, yet it can still leave a symmetrical pattern of magnetized rock. The molten material emerging from a spreading center becomes magnetized because as it cools it captures the prevailing direction of the earth's periodically changing magnetic field. In the instance illustrated here the right-hand plate moves out to the right while the left-hand plate remains stationary. In *1* hot material from a dike arrives at the surface, cools and splits down the middle. In *2* the next injection of material arrives in the crevice between the two halves of the preceding mass of rock. The new mass is therefore half the width of the preceding mass farther from the stationary plate than the preceding mass of material itself was. In *3* the new material has cooled and split in its turn and another mass has appeared that is a whole width farther from the left-hand plate. As long as the center moves at half the speed at which the right-hand plate moves away the magnetic bands remain symmetrical. If plate moved faster or more slowly, they would be jumbled.

material dissolves, and only red clay and other resistant components remain.

For reasons only partly known the sediment is not uniformly distributed. At the spreading center the newly created crust is of course bare of sediment, and within 100 kilometers of such a center the calcareous ooze is rarely thick enough to measure. The ooze accumulates at an average rate of 10 meters for each million years, during which time the plate moves horizontally from 10,000 to 100,000 meters and sinks 100 meters. Where the red clay appears, it accumulates at a rate of less than one meter per million years.

The puzzle deepens when one considers that sediment on oceanic crust older than 20 million years stops increasing in thickness after it sinks to the depth where the calcareous ooze dissolves. Indeed, in many places the age of the oldest sediment is about the same as the volcanic layer on which it lies. It would therefore seem that almost all the deep-ocean sediment accumulates in narrow zones on the flanks of the mid-ocean ridges. If this is correct, it has yet to be explained.

The volcanic layer forms mainly at the spreading center. Volcanoes and vents on the slopes of the mid-ocean ridge contribute a certain amount of oceanic tholeiite to it. It can be said in general that the thickness of the volcanic layer decreases as the spreading rate increases. If the crust spreads slowly, the material has time to accumulate. Fast spreading reduces this time and therefore the accumulation. The conclusion can be drawn that the rate at which the volcanic-layer material is discharged is nearly constant. These relations are based on only 10 observations, but they apply to spreading rates from 1.4 centimeters to 12 centimeters per year, and to thicknesses from .8 kilometer to 3.8 kilometers.

The total flow of volcanic material from all active spreading centers is about four cubic kilometers per year—four times the flow on the continents. Not the slightest sign of this volcanism on the ocean bottom can be detected at the surface of the ocean, with one possible exception: late in the 19th century a ship reported seeing smoke rising from the waters above the equatorial Mid-Atlantic Ridge. The British oceanographer Sir John Murray remarked that he hoped the smoke signified the emergence of an island, since the Royal Navy needed a coaling station at that point.

Like the volcanic layer, the oceanic

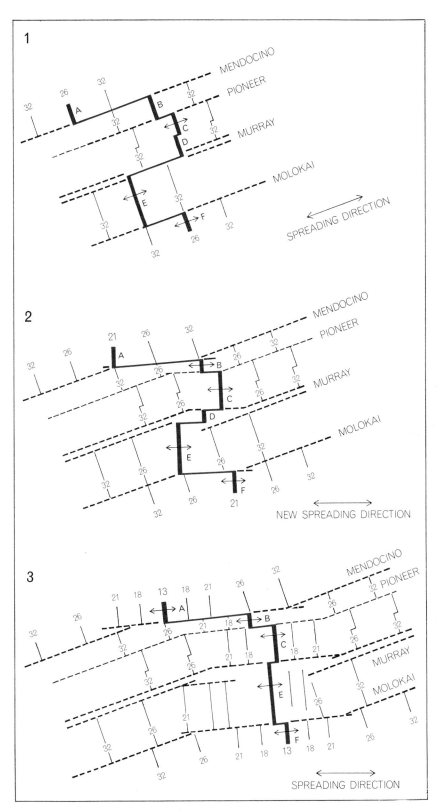

HOW A PLATE MOVES is revealed by the patterns of magnetic bands (time of formation is indicated by numbers) and by the relation between ridges and faults in the northeastern Pacific. At the time illustrated in *1* material from the Murray fault and other faults connected segments of the ridge (*indicated by letters*) offset from one another by plate motion. At the time shown in *2* the spreading direction changed. The readjustment of the plates has shortened the Pioneer-Mendocino ridge segment (*B*) while lengthening and reorienting the Pioneer, Mendocino and Molokai transform faults. In *3* the Mendocino fault remains the same length, but between most of the other ridges faults have been shortened or have almost disappeared as ridge segments tended to rejoin one another. Between Murray and Molokai faults ridge has jumped eastward, and one segment (*D*) has vanished.

layer forms at the spreading center. Acoustical measurements of the thickness of this layer at spreading centers, on the flanks of mid-ocean ridges and on the deep-ocean floor show, however, that at least part of the oceanic layer evolves slowly from the mantle rather than solidifying quickly and completely at the spreading center. At the spreading center the thickness of the layer depends on how fast the ocean floor moves. In regions such as the South Atlantic, where the floor spreads at a rate of two centimeters per year, no oceanic layer forms within a few hundred kilometers of the spreading center. Farther away from the spreading center the oceanic layer accumulates rapidly, reaching a normal thickness of four to five kilometers on the flank of the mid-ocean ridge. A spreading rate of three centimeters per year is associated with an oceanic layer roughly two kilometers deep at the center that thickens by one kilometer in 13 million years. A plate with a spreading rate of eight centimeters per year is three kilometers deep at the center and thickens by one kilometer in 20 million years. Thus the thinner the initial crust is, the faster the thickness increases as the crust spreads.

As a plate flows continuously from the spreading center, faulting, volcanic eruptions and lava flows along the length of the mid-ocean ridge build its mountains and escarpments. This process can be most easily observed in Iceland, a part of the Mid-Atlantic Ridge that has grown so rapidly it has emerged from the ocean. A central rift, 45 kilometers wide at its northern end, cuts the island parallel to the ridge. The sides of the rift consist of active, steplike faults. There are other step faults on the rift floor, which is otherwise dominated by a large number of longitudinal fissures. Some of these fissures are open and filled with dikes. Fluid lava wells up from the fissures and either buries the surrounding mountains, valleys and faults or forms long, low "shield" volcanoes. Two hundred such young volcanoes, which have been erupting about once every five years over the past 1,000 years, dot the floor of the rift. Thirty of them are currently active.

Just as a balance between spreading and sinking shapes the slopes of the mid-ocean ridges, so does a balance between lava discharge and spreading build undersea mountains, hills and valleys. High mountains normally form at slow centers where spreading proceeds at two to 3.5 centimeters per year. In contrast, a spreading center that opens at a rate of five to 12 centimeters per year produces long hills less than 500 meters high. This relationship is a natural consequence of the long-term constancy of the lava discharge. Over a short period of time, however, the lava discharge may fluctuate or pulsate, a picture suggested by the fact that the thick volcanic layer associated with slow spreading can consist of volcanic mountains (which represent copious flow) separated by valleys covered by a thinner volcanic layer.

The volcanic activity and faulting that first appear near the spreading centers decrease rapidly as the plate ages and material moves toward its center, but volcanic activity in some form is never entirely absent. Small conical volcanoes are found on crust only a few hundred thousand years old near spreading centers, and active circular volcanoes such as those of Hawaii exist even in the middle of a plate. It would appear that the great cracks that serve as conduits for dikes and lava flows are soon sealed as a plate ages and spreads. Volcanic activity is then concentrated in a few central vents, created at different times and places.

Many of these vents remain open for tens of millions of years, judging by the size and distribution of the different classes of marine volcanoes. First, the biggest volcanoes are increasingly big at greater distances from a spreading center, which means they must continue to erupt and grow as the crust ages and sinks, even when the age of the crust exceeds 10 million years. In most places, in fact, a volcano needs at least 10 million years in order to grow large enough to become an island. Volcanoes that discharge lava at a rate lower than 100 cubic kilometers per million years never become islands because the sea floor sinks too fast for them to reach the sea surface.

Other volcanoes drifting with a spreading ocean floor may remain active or become active on crust that is 100 million years old (as the volcanoes of the Canary Islands have). Normally, however, volcanoes become inactive by the time the crust is 20 to 30 million years old. This is demonstrated by the existence of guyots, drowned ancient island volcanoes that were submerged by the gradual sinking of the aging crust. Guyots are found almost entirely on crust that is more than 30 million years old, such as the floor of the western equatorial Pacific.

Traditionally it has been thought that marine volcanoes spew lava from a magma chamber located deep in the mantle. Some volcanoes have a top composed of alkali basalt, slopes with transitional basalt outcrops and a base of oceanic tholeiite, and it was therefore assumed that the lava in the magma chamber became differentiated into components that, rather like a pousse-café, separated into several layers of different

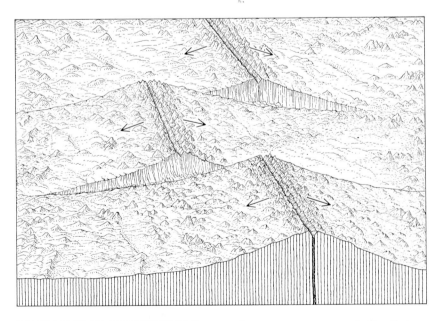

RIDGE-RIDGE TRANSFORM FAULT appears between two segments of ridge that are displaced from each other. Mountains are built, earthquakes shake the plate edges and volcanoes erupt in such an area because of the forces generated as plates, formed at the spreading centers under the ridges, slide past each other in opposite directions. On outer slopes of the mid-ocean ridges, however, this intense seismic activity appears to subside.

kinds of material, each of which followed the layer above it up the spout. The emergence of plate tectonics and continental drift as respectable concepts have now brought this view of volcanic action into question.

It remains perfectly possible for a volcano to drift for tens of millions of years over hundreds of kilometers while tapping a single magma chamber embedded deep in the mantle. The motion of the plates, however, suggests another hypothesis. According to this view, the volcano and its conduit drift along with the crust as the conduit continually taps different parts of a relatively stationary magma that is ready to yield various kinds of lava whenever a conduit appears. In actuality the composition of the lava usually changes only slightly after the first 10 to 20 kilometers of drifting. Although the older hypothesis is still reasonable, the newer one must also be considered because it explains the facts equally well.

In addition to their characteristic volcanoes and mountains, spreading centers are marked by median valleys, which in places such as the North Atlantic or the northwestern Indian Ocean are deeper than the surrounding region. These rifts are commonly found in centers opening at a rate of two to five centimeters per year. The deepest rifts, which may go as deep as 1,000 to 1,300 meters below the surrounding floor, are associated with spreading at three to four centimeters per year. Only one valley is known to be associated with spreading at five to 12 centimeters per year. Although rifts are not found in all spreading centers, they usually do appear in conjunction with a slow center. Both of these features are also associated with volcanic activity.

The mid-ocean ridges, as we have noted, seldom run unbroken for more than a few hundred kilometers. They are interrupted by fracture zones, and the segments are shifted out of line with respect to one another. These fracture zones run at right angles to the ridge and connect the segments. Where they lie between the segments they are termed ridge-ridge transform faults, which are the site of intense geological activity. As the two edges of the fault slide past each other they rub and produce earthquakes. The slope of a transform fault drops steeply from the crest of one ridge segment to a point halfway between it and the adjacent segment and then climbs to the top of the adjacent segment, reflecting the fact that the crust is elevated at the spreading center and subsides at some distance from it [*see illustration on page 84*].

Like spreading centers, fracture zones have their own complex geology. In these

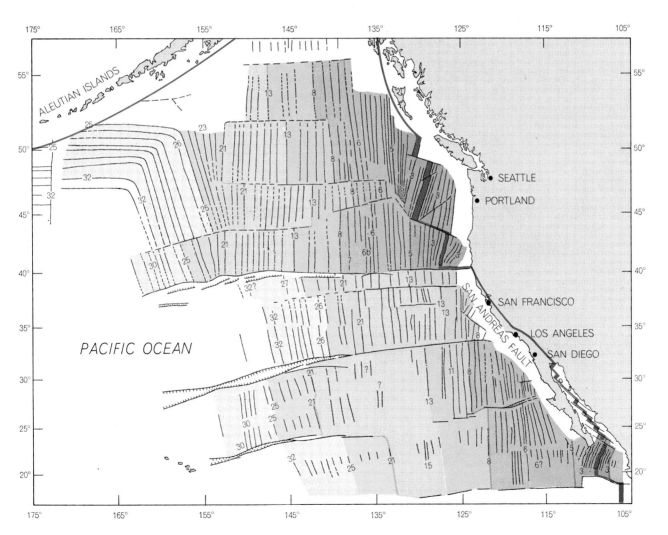

MAGNETIC PATTERNS reveal how the plate forming the floor of the northeastern Pacific has moved. Its active eastern edge now stretches from Alaska through California (where it forms the San Andreas fault) to the Gulf of California. In the gulf spreading centers break into short segments joined by active faults. Plate motion is opening the gulf and moving coastal California in the direction of the Aleutians. To the south lies the Great Magnetic Bight, formed by three plates that spread away from one another.

areas the ridges stand as much as several kilometers high, and the troughs are equally deep. It appears that the same volcanic forces that shape the main ridges produce the mountains and valleys of the faults. As fracture zones open they slowly leak lava from hot dikes. At the same time the crust sinks away from the fault line, and this balance produces high mountains.

Beyond the spreading centers the fracture zones become the inactive remains of earlier faulting. The different rates at which these outer flanks of the mid-ocean ridge sink do produce some vertical motion as the scarps of the fracture zone decay. This may account for the few earthquakes in these areas. I should emphasize that it is not known if horizontal motion is also absent from such dead fracture zones. It is not necessary, however, to postulate such motion in order to explain existing observations.

A fracture zone can become active again at any time, but if it does so, it becomes the side of a smaller new plate rather than part of the trailing edge of an old one. If the flank fracture zones are as quiescent as they appear to be, then the plates forming the earth's crust are large and long-lived. If these fracture zones were active, on the other hand, it could only be concluded that each one marked the flank of a small, elongated plate.

The direction the plate is moving can be deduced from the magnetic pattern that runs at right angles to the fractures in the fracture zone. When the direction of plate motion changes, the direction in which the fracture zone moves also changes. This change in direction can be most clearly seen in the northeastern Pacific, where our knowledge is most detailed. On this part of the ocean floor the changes of direction have taken place at the same time in many zones, indicating that the entire North Pacific

plate has change direction as a unit [see *illustration on page 85*].

On the bottom of the Pacific and the North Atlantic the magnetic patterns are sometimes garbled. Old transform faults may have vanished if short segments of spreading center have been united by reorientation. By the same token new transform faults may have formed if the change in plate motion has been too rapid to be accommodated by existing motions. Thus fracture zones may be discontinuous. They may start and stop abruptly, and the offset of the magnetic patterns may change from place to place along them without indicating any activity except at the former edges of plates.

Some patterns are even harder to interpret. Douglas J. Elvers and his colleagues in the U.S. Coast and Geodetic Survey discovered an abrupt boomerang-shaped bend in the magnetic pattern south of the Aleutians. The arms of this

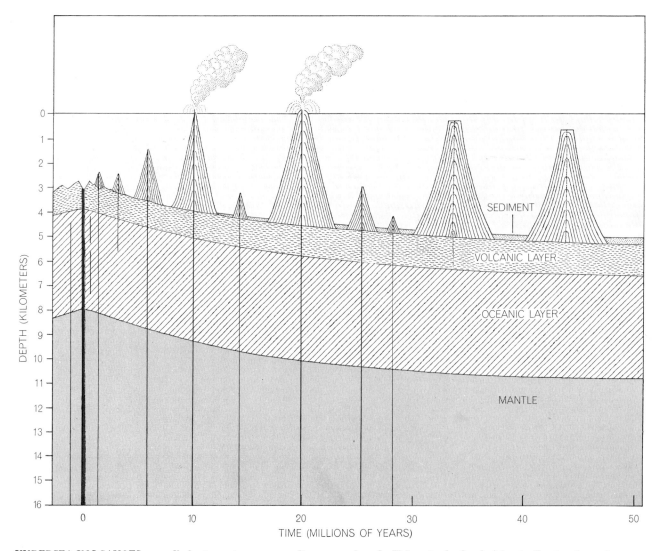

UNDERSEA VOLCANOES normally begin to rise near spreading centers. Then they ride along on the moving plate as they grow. If a volcano rises fast enough to surmount the original depth of the water and the sinking of the ocean floor, it emerges as an island such as St. Helena in the South Atlantic. To rise above the water an undersea volcano must grow to a height of about four kilometers in 10 million years. Island volcanoes sink after 20 to 30 million years and become the sediment-capped seamounts called guyots.

configuration, which Elvers calls the Great Magnetic Bight, are offset by fracture zones at right angles to the magnetic pattern. The Great Magnetic Bight seems to have required impossible forms of sea-floor spreading and plate movement. However, Walter C. Pitman III and Dennis E. Hayes of the Lamont-Doherty Geological Observatory, among others, have been able to show that the configuration is fully understandable if it is assumed that the trailing edges of three plates met and formed a Y. Similar complex patterns have now been found in the Atlantic and the Pacific. Indeed, if the transform faults are perpendicular to the spreading centers and two spreading rates are known, both the orientation and the spreading rate of the third spreading center can be calculated even before it is mapped. One can also calculate the orientation and spreading rate for a third center that has already vanished in a trench.

As a crustal plate grows, its leading edge is destroyed at an equal rate. Sometimes this edge slides under the oncoming edge of another plate and returns to the asthenosphere. When this happens, a deep trench such as the Mariana Trench and the Tonga Trench in the Pacific is formed. In other areas the movement of the crust creates young mountain ranges. Xavier Le Pichon of the Lamont-Doherty Geological Observatory concludes that the occurrence of one event or the other is a function of the rate at which plates are moving together. If the rate is less than five to six centimeters per year, the crust can absorb the compression and buckles up into large mountain ranges such as the Himalayas. In these ranges folding and overthrusting deform and shorten the crust. If the rate is higher, the plate breaks free and sinks into the mantle, creating an oceanic trench in which the topography and surface structure indicate tension.

Several crustal sinks are no longer active. Their past, however, can be deduced from their geology. Large-scale folding and thrust-faulting can be taken as evidence of the former presence of a crustal sink, although such deformation can also arise in other ways. Certain types of rock may also indicate the formation of a trench. The arcs of islands that lie parallel to trenches, for instance, are characterized by volcanoes that produce andesitic lavas, which are quite different from the basaltic lavas of the ocean floor. The trenches themselves, and the deep-sea floor in general, are featured by deposits of graywackes and cherts. These rocks are commonly found exposed on land in the thick prisms of

sediment that lie in geosynclines: large depressed regions created by horizontal forces resembling those generated by a drifting crustal plate. Thus the presence of some or all of these types of rock may indicate the former existence of a crustal sink. This linking of marine geology at spreading centers with land geology at crustal sinks is becoming one of the most fruitful aspects of plate tectonics. Still, crustal sinks are by no means as informative about the history of the ocean floor as spreading centers are, because in such sinks much of the evidence of past events is destroyed. Even if the leading edge of a plate was once the side of a plate (or vice versa), there would be no way to tell them apart.

At the boundary between the land and the sea a puzzle presents itself. The sides of an oceanic trench move together at more than five centimeters per year, and it would seem that the sediment sliding into the bottom of the trench should be folded into pronounced ridges and valleys. Yet virtually undeformed sediments have been mapped in trenches by David William Scholl and his colleagues at the U.S. Naval Electronics Laboratory Center. Furthermore, the enormous quantity of deep-ocean sediment that has presumably been swept up to the margins of trenches cannot be detected on sub-bottom profiling records. There are many ingenious (but unpublished) explanations of the phenomenon in terms of plate tectonics. One of them may conceivably be correct. According to that hypothesis, the sediments are intricately folded in such a way that the slopes and walls of trenches cannot be detected by normal survey techniques, which look at the sediments from the ocean surface and along profiles perpendicular to the slopes. This kind of folding could be detected only by trawling a recording instrument across the trench much closer to the bottom or by crossing the slope at an acute angle.

The concepts of sea-floor spreading and plate tectonics allow a quantitative evaluation of the interaction of many important variables in marine geology. By combining empirical observation with theory it is possible not only to explain but also to predict the thickness and age of sediments in a given locality, the scale and orientation of topographic relief, the thickness of various crustal layers, the orientation and offsetting of magnetic patterns, the distribution and depth of drowned ancient islands, the occurrence of trenches and young mountain ranges, the characteristics of earthquakes, and many other previously unrelated and un-

predictable phenomena. This revolution in marine geology may take some years to run its course. Ideas are changing, and new puzzles present themselves even as the old ones are solved. The only certainty is that the subject will never be the same again.

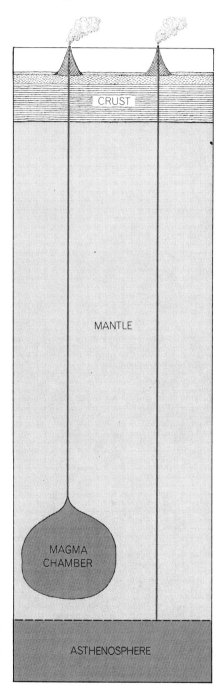

UNDERSEA ERUPTIONS can be explained in two ways, both consistent with observed facts. Since a volcano consists of different kinds of rock, it was originally thought that its conduit carried different forms of lava up from a magma chamber 50 kilometers or more down in the mantle. Now that crustal plates have been found to move, another theory must be considered. According to this idea, the conduit reaches through the mantle and taps several different kinds of magma at different places in the asthenosphere.

10

THE ORIGIN OF THE OCEANS

SIR EDWARD BULLARD
September 1969

The earth is uniquely favored among the planets: it has rain, rivers and seas. The large planets (Jupiter, Saturn, Uranus and Neptune) have only a small solid core, presumably overlain by gases liquefied by pressure; they are also surrounded by enormous atmospheres. The inner planets are more like the earth. Mercury, however, has practically no atmosphere and the side of the planet facing the sun is hot enough to melt lead. Venus has a thick atmosphere containing little water and a surface that, according to recent measurements, may be even hotter than the surface of Mercury. Mars and the moon appear to show us their primeval surfaces, affected only by craters formed by the impact of meteorites, and perhaps by volcanoes. Only on the earth has the repetition of erosion and sedimentation —"the colossal hour glass of rock destruction and rock formation"—run its course cycle after cycle and produced the diverse surface that we see. The mountains are raised and then worn away by falling and running water; the debris is carried onto the lowlands and then out to the ocean. Geologically speaking, the process is rapid. The great plateau of Africa is reduced by a foot in a few thousand years, and in a few million years it will be near sea level, like the Precambrian rocks of Canada and Finland. All trace of the original surface of the earth has been removed, but as far back as one can see there is evidence in rounded, water-worn pebbles for the existence of running water and therefore, presumably, of an ocean and of dry land.

The obvious things that no one comments on are often the most remarkable; one of them is the constancy of the total volume of water through the ages. The level of the sea, of course, has varied from time to time. During the ice ages, when much water was locked up in ice sheets on the continents, the level of the sea was lower than it is at present, and the continental shelves of Europe and North America were laid bare. Often the sea has advanced over the coastal plains, but never has it covered all the land or even most of it. The mechanism of this equilibrium is unknown; it might have been expected that water would be expelled gradually from the interior of the earth and that the seas would grow steadily larger, or that water would be dissociated into hydrogen and oxygen in the upper atmosphere and that the hydrogen would escape, leading to a gradual drying up of the seas. These things either do not happen or they balance each other.

The mystery is deepened by the almost complete loss of neon from the earth; in the sun and the stars neon is only a little rarer than oxygen. The neon was presumably lost when the earth was built up from dust and solid grains because neon normally does not form compounds, but if that is so, why was the water not lost too? Water has a molecular weight of 18, which is less than the atomic weight of neon, and thus should escape more easily. It looks as if the water must have been tied up in compounds, perhaps hydrated silicates, until the earth had formed and the neon had escaped. Water must then have been released as a liquid sometime during the first billion years of the earth's history, for which we have no geological record. The planet Mercury and the moon would have been too small to retain water after it was released. Mars seems to have been able to retain a trace, not enough to make oceans but enough to be detectable by spectroscopy.

These speculations about the early history of the earth are open to many doubts. The evidence is almost non-existent, and all one can say is, "It might have been that...." The great increase in understanding of the present state and recent history of the ocean basins that we have gained in the past 20 years is something quite different. For the first time the geology of the oceans has been studied with energy and resources commensurate with the tremendous task. It turns out that the main processes of geology can be understood only when the oceans have been studied; no amount of effort on land could have told us what we now know. The study of marine geology has unlocked the history of the oceans, and it seems likely to make intelligible the history of the continents as well. We are in the middle of a rejuvenating process in geology comparable to the one that physics experienced in the 1890's and to the one that is now in progress in molecular biology.

The critical step was the realization that the oceans are quite different from the continents. The mountains of the oceans are nothing like the Alps or the Rockies, which are largely built from folded sediments. There is a world-encircling mountain range—the mid-ocean ridge—on the sea bottom, but it is built entirely of igneous rocks, of basalts that have emerged from the interior of the earth. Although the undersea mountains have a covering of sediments in

RED SEA and the Gulf of Aden represent two of the newest seaways created by the worldwide spreading of the ocean floor. In this photograph, taken at an altitude of 390 miles from the spacecraft *Gemini 11* in September, 1966, the Red Sea separates Ethiopia *(at left)* from the Arabian peninsula *(at right)*. The Gulf of Aden lies between the southern shore of Arabia and Somalia. The excellent fit between the drifting land masses is depicted in the illustrations on page 91.

many places, they are not made of sediments, they are not folded and they have not been compressed.

A cracklike valley runs along the crest of the mid-ocean ridge for most of its length, and it is here that new ocean floor is being formed today [see illustration on next two pages]. From a study of the numerous earthquakes along this crack it is clear that the two sides are moving apart and that the crack would continually widen if it were not being filled with material from below. As the rocks on the two sides move away and new rock solidifies in the crack, the events are recorded by a kind of geological tape recorder: the newly solidified rock is magnetized in the direction of the earth's magnetic field. For at least the past 10,000 years, and possibly for as long as 700,000 years, the north magnetic pole has been close to its present location, so that the magnetic field is to the north and downward in the Northern Hemisphere, and to the north and upward in the Southern Hemisphere. As the cracking and the spreading of the ocean floor go on, a strip of magnetized rock is produced. Then one day, or rather in the course of several thousand years, the earth's field reverses, the next effusion of lava is magnetized in the reverse direction and a strip of reversely magnetized rocks is built up between the two split halves of the earlier strip. The reversals succeed one another at widely varying intervals; sometimes the change comes after 50,000 years, often there is no change for a million years and occasionally, as during the Permian period, there is no reversal for 20 million years. The sequence of reversals and the progress of spreading is recorded in all the oceans by the magnetization of the rocks of the ocean floor. The message can be read by a magnetometer towed behind a ship.

We now have enough examples of these magnetic messages to leave no doubt about what is happening. It is a truly remarkable fact that the results of magnetic surveys in the South Pacific can be explained—indeed predicted—from the sequence of reversals of the direction of the earth's magnetic field known from magnetic and age measurements, made quite independently on lavas in California, Africa and elsewhere. The only adjustable factor in the calculation is the rate of spreading. Such worldwide theoretical ideas and such detailed agreement between calculation and theory are rare in geology, where theories are usually qualitative, local and of little predictive value.

The speed of spreading on each side of a mid-ocean ridge varies from less than a centimeter per year to as much as eight centimeters. The fastest rate is the one from the East Pacific Rise and the slowest rates are those from the Mid-Atlantic Ridge and from the Carlsberg Ridge of the northwest Indian Ocean. The rate of production of new terrestrial crust at the central valley of a ridge is the sum of the rates of spreading on the two sides. Since the rates on the two sides are commonly almost equal, this sum is twice the rate on each side and may be as much as 16 centimeters (six inches) per year. Such rates are, geologically speaking, fast. At 16 centimeters per year the entire floor of the Pacific Ocean, which is about 15,000 kilometers (10,000 miles) wide, could be produced in 100 million years.

When the mid-ocean ridges are examined in more detail, they are found not to be continuous but to be cut into sections by "fracture zones" [see top illustration on page 94]. A study of the earthquakes on these fracture zones shows that the separate pieces of ridge crest on the two sides of a fracture zone are not moving apart, as might seem likely on first consideration. The two pieces of ridge remain fixed with respect to each other while on each side a plate of the crust moves away as a rigid body; such a fracture is called a transform fault. The earthquakes occur only on the piece of the fracture zone between the two ridge crests; there is no relative motion along the parts outside this section.

If two rigid plates on a sphere are spreading out on each side of a ridge that is crossed by fracture zones, the relative motion of the two plates must be a rotation around some point, termed the pole of spreading. The "axis of spreading," around which the rotation takes place, passes through this pole and the center of the earth. The existence of a pole of spreading and an axis of spreading is geometrically necessary, as was shown by Leonhard Euler in the

PROBABLE ARRANGEMENT of continents before the formation of the Atlantic Ocean was determined by the author with the aid of a computer. The fit was made not at the present coastlines but at the true edge of each continent, the line where the continental shelf (dark brown) slopes down steeply to the sea floor. Overlapping land and shelf areas are reddish orange; gaps where the continental edges do not quite meet are dark blue. At present the entire western Atlantic is moving as one great plate carrying both North America and South America with it. At an earlier period the two continents must have moved independently.

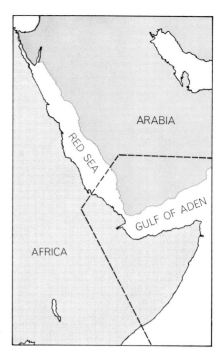

RUPTURE OF MIDDLE EAST is being caused by the widening of the Red Sea and the Gulf of Aden. Some 20 million years ago the Arabian peninsula was joined to Africa, as evidenced by the remarkable fit between shorelines (see illustration below). The area within the *Gemini 11* photograph on page 89 is shown by the broken lines.

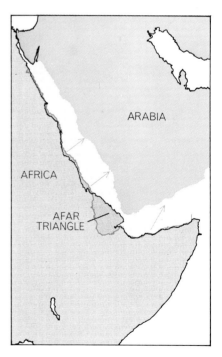

FIT OF SHORELINES of Arabia and Africa works out most successfully if the African coast (black) is left intact and if the Arabian coast (color) is superposed in two separate sections. In the reconstruction a corner of Arabia overlaps the "Afar triangle" in northern Ethiopia, an area that now has some of the characteristics of an ocean floor.

18th century. If the only motion on the fracture zones is the sliding of the two plates past each other, then the fracture zones must lie along circles of latitude with respect to the pole of spreading, and the rates of spreading at any point on the ridge must be proportional to the perpendicular distance from the point to the axis of spreading [*see bottom illustration on page 94*].

All of this is well verified for the spreading that is going on today. The rates of spreading can be obtained from the magnetic patterns and the dates of the reversals. The poles of spreading can be found from the directions of the fracture zones and checked by the direction of earthquake motions. It turns out that the ridge axes and the magnetic pattern are usually almost at right angles to the fracture zones. This is not a geometrical

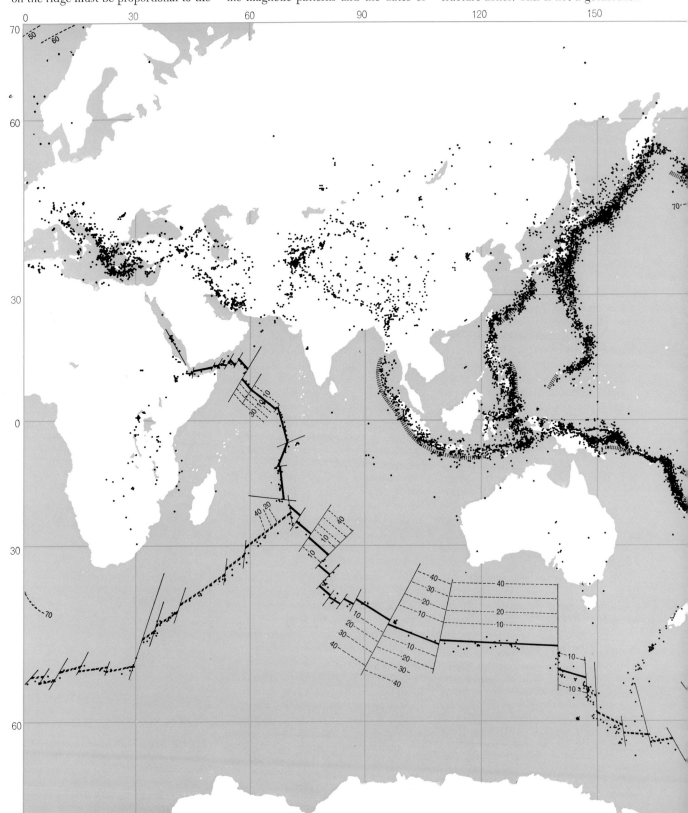

OCEANIC GEOLOGY has turned out to be much simpler than the geology of the continents. New ocean bottom is continuously being extruded along the crest of a worldwide system of ridges (*thick black lines*). The present position of material extruded at intervals of 10 million years, as determined by magnetic studies, appears as broken lines parallel to the ridge system, which is offset by fracture

necessity, but when it does happen it means that the lines of the ridge axes and of the magnetic pattern must, if they are extrapolated, go through the pole of spreading. If the ridge consists of a number of offset sections at right angles to the fracture zones, the axes of these sections will converge on the pole of spreading. It is one of the surprises of the work at sea that this rather simple geometry embraces so large a part of the facts. It seems that marine geology is truly simpler than continental geology and that this is not merely an illusion based on our lesser knowledge of the oceans.

The regularity of the magnetic pattern suggests that the ocean floor can move as a rigid plate over areas several thousand kilometers across. The thickness of the rigid moving plate is quite uncertain,

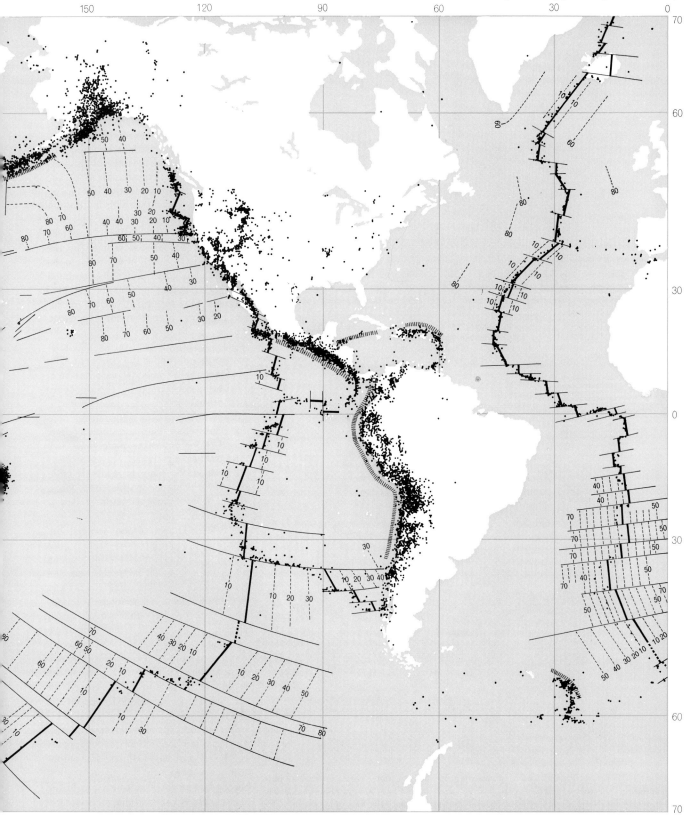

zones (*thin black lines*). Earthquakes (*black dots*) occur along the crests of ridges, on parts of the fracture zone and along deep trenches. These trenches, where the ocean floor dips steeply, are represented by hatched bands. At the maximum estimated rate of sea-floor spreading, about 16 centimeters a year, the entire floor of the Pacific Ocean could be created in perhaps 100 million years.

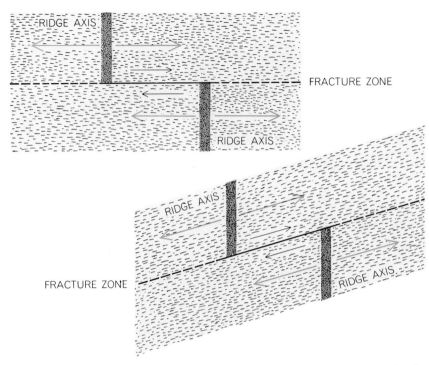

MOTION AT AXIS OF RIDGE consists of an opening of an axial crack (*vertical bands*) where two plates separate (*arrows*). Often the ridge is offset by a fracture zone, making a transform fault where one plate slips past another. The motion must be parallel to the fracture zone. It is usually at right angles to the ridge (*upper left*) but need not be (*lower right*).

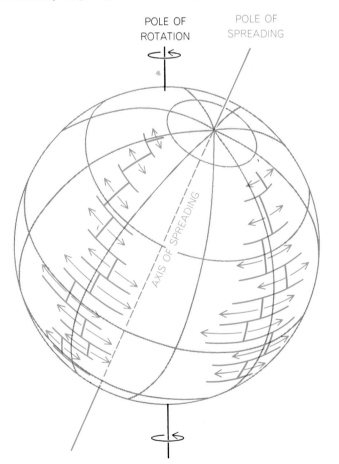

MOTION OF RIGID PLATES on a sphere requires that the plates rotate around a "pole of spreading" through which passes an "axis of spreading." Plates always move parallel to the fracture zones and along circles of latitude perpendicular to the axis of spreading. The rate of spreading is slowest near the pole of spreading and fastest 90 degrees away from it. The spreading pole can be quite remote from the sphere's pole of rotation.

but a value between 70 and 100 kilometers seems likely. If this is so, the greater part of the plate will be made of the same material as the upper part of the earth's mantle—probably of peridotite, a rock largely composed of olivine, a silicate of magnesium and iron, $(Mg, Fe)_2$-SiO_4. The basaltic rocks of the oceanic crust will form the upper five kilometers or so of the plate, with a veneer of sediments on top.

What happens at the boundary of an ocean and a continent? Sometimes, as in the South Atlantic, nothing happens; there are no earthquakes, no distortion, nothing to indicate relative motion between the sea floor and the bordering continent. The continent can then be regarded as part of the same plate as the adjacent ocean floor; the rocks of the continental crust evidently ride on top of the plate and move with it. In other places there is another kind of coast, what Eduard Suess called "a Pacific coast." Such a coast is typified by the Pacific coast of South America. Here the oceanic plate dives under the continent and goes down at an angle of about 45 degrees. On the upper surface of this sloping plate there are numerous earthquakes—quite shallow ones near the coast and others as deep as 700 kilometers inland, under the continent. The evidence for the sinking plate has been beautifully confirmed by the discovery that seismic waves from shallow earthquakes and explosions, occurring near the place where the plate starts its dive, travel faster down the plate than they do in other directions. This is expected because the plate is relatively cold, whereas the upper mantle, into which the plate is sinking, is made of similar material but is hot.

Little is known of the detailed behavior of the plate; further study is vital for an understanding of the phenomena along the edges of continents. Near the point where the plate turns down there is an ocean deep, whose mode of formation is not precisely understood, but if a plate goes down, it is not difficult to imagine ways in which it could leave a depression in the sea floor. It is probable that, as the plate goes down, some of the sediment on its surface is scraped off and piled up in a jumbled mass on the landward side of the ocean deep. This sediment may later be incorporated in the mountain range that usually appears on the edge of the continent. The mountain range bordering the continent commonly has a row of volcanoes, as in the Andes. The lavas from the volcanoes are frequently composed of andesites, which are different from the lavas of the mid-ocean ridges in that they contain more silica. It may reasonably be supposed that they are formed by the partial melting of the descending plate at a depth of about 150 kilometers. The first material to melt will contain more silica than the remaining material; it is also possible that the melted material is contaminated by granite as it rises to the surface through the continental rocks.

In many places the sinking plate goes down under a chain of islands and not under the continent itself. This happens in the Aleutians, to the south of Indonesia, off the islands of the Tonga group, in the Caribbean and in many other places. The volcanoes are then on the islands and the deep earthquakes occur under the almost enclosed sea behind the chain or arc of islands, as they do in the Sea of Japan, the Sea of Okhotsk and the Java Sea.

The destruction of oceanic crust explains one of the great paradoxes of geology. There have always been oceans, but the present oceans contain no sediment more than 150 million years old and very little sediment older than 80 million years. The explanation is that the older sediments have been carried away with the plates and are either piled up at the edge of a continent or are carried down with a sinking plate and lost in the mantle.

The picture is simple: the greater part of the earth's surface is divided into six plates [*see illustration on next two pages*]. These plates move as rigid bodies, new material for them being produced from the upper mantle by lava emerging from the crack along the crest of a mid-ocean ridge. Plates are destroyed at the oceanic trenches by plunging into the mantle, where ultimately they are mixed again with the material whence they came. The scheme is not yet established in all its details. Perhaps the greatest uncertainty is in the section of the ridge running south of South Africa; it is not clear how much of this is truly a ridge and a source of new crust and how much is a series of transform faults with only tangential motion. It is also uncertain whether the American and Eurasian plates meet in Alaska or in Siberia. It appears certain, however, that they do not meet along the Bering Strait.

A close look at the system of ridges, fracture zones, trenches and earthquakes reveals many other features of great interest, which can only be mentioned here. The Red Sea and the Gulf of Aden appear to be embryo oceans [*see illustration on page 89*]. Their floors are truly oceanic, with no continental rocks; along their axes one can find offset lengths of crack joined by fracture zones, and magnetic surveys show the worldwide magnetic pattern but only the most recent parts of it. These seas are being formed by the movement of Africa and Arabia away from each other. A detailed study of the geology, the topography and the present motion suggests that the separation started 20 million years ago in the Miocene period and that it is still continuing. If this is so, there must have been a sliding movement along the Jordan rift valley, with the area to the east having moved about 100 kilometers northward with respect to the western portion. There must also have been an opening of the East African rift valley by 65 kilometers or so.

The first of these displacements is well established by geological comparisons between the two sides of the valley, and it should be possible to verify the second. The reassembly of the pieces requires that the southwest corner of Arabia overlap the "Afar triangle" in northern Ethiopia [*see bottom illustration on page 91*]. This area should therefore be part of the embryo ocean. The fact that it is dry land presented a substantial puzzle, but recently it has been shown that the oceanic magnetic pattern extends over the area; it is the only land area in the world where this is known to happen. It seems likely that the Afar triangle is in some sense oceanic. The results of gravity surveys, seismic measurements and drilling will be awaited with interest. On this picture Arabia and the area to the north comprise a small plate separate from the African and Asian plates. The northern boundary of this small plate may be in the mountains of Iran and Turkey, where motion is proceeding today.

A number of other small plates are known. There is one between the Pacific coast of Canada and the ridge off Vancouver Island; it is probable that this is being crumpled at the coast rather than diving under the continent. Farther south the plate and the ridge from which it spread may have been overrun by the westward motion of North America. The ridge appears again in the Gulf of California, which is similar in many ways to the Red Sea and the Gulf of Aden. From the mouth of the Gulf of California the ridge runs southward and is joined by an east-west ridge running through the Galápagos Islands. The sea floor bounded by the two ridges and the trench off Central America seems to constitute a separate small plate.

For the past four million years we can date the lavas on land with enough

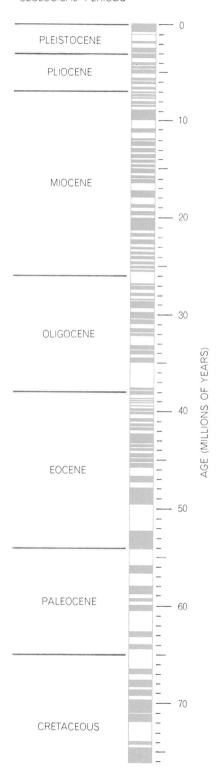

GEOLOGICAL PERIODS

PLEISTOCENE

PLIOCENE

MIOCENE

OLIGOCENE

EOCENE

PALEOCENE

CRETACEOUS

0

10

20

30

40

50

60

70

AGE (MILLIONS OF YEARS)

REVERSALS of the earth's magnetic field can be traced back more than 70 million years using magnetic patterns observed on the sea floor. The timetable of reversals for the most recent four million years was obtained by dating reversals in lava flows on land. Extrapolations beyond that assume that the sea floor spread at a constant rate. Colored bars show periods when the direction of the magnetic field was as it is now.

accuracy to give a timetable of magnetic reversals that can be correlated with the magnetic pattern on the sea bottom. For this period the rates of spreading from the ridges have remained constant. Further back we have a long series of reversals recorded in the ocean floor, but we cannot date them by comparison with lavas on land because the accuracy of the dates is insufficient to put the lavas in order. A rough guess can be made of the time since the oldest part of the magnetic pattern was formed by assuming that the rates have always been what they are today. This yields about 70 million years in the eastern Pacific and the South Atlantic. In fact the spacings of the older magnetic lineations are not in a constant proportion in the different oceans. The rates of spreading must therefore vary with time when long periods are considered. Directions of motion have also changed during this

period, as can be seen from the departure of the older parts of some of the Pacific fracture zones from circles of latitude around the present pole of spreading. A change of direction is also shown by the accurate geometrical and geochronological fit that can be made between South America and Africa [see illustration on page 90]. A rotation around the present pole of spreading will not bring the continents together; it is therefore likely that in the early stages of the separation motion was around a point farther to the south.

The ideas of the development of the earth's surface by plate formation, plate motion and plate destruction can be checked with some rigor by drilling. If they are correct, drilling at any point should show sediments of all ages from the present to the time at which this part of the plate was in the central valley of

the ridge. Under these sediments there should be lavas of about the same age as the lowest sediments. From preliminary reports of the drilling by the JOIDES project (a joint enterprise of five American universities) it seems that this expectation has been brilliantly verified and that the rate of spreading has been roughly constant for 70 million years in the South Atlantic. Such studies are of great importance because they will give firm dates for the entire magnetic pattern and provide a detailed chronology for all parts of the ocean floor.

The process of consumption of oceanic crust at the edge of a continent may proceed for tens of millions of years, but if the plate that is being consumed carries a continental fragment, then the consumption must stop when the fragment reaches the trench and collides with the continent beyond it. Because

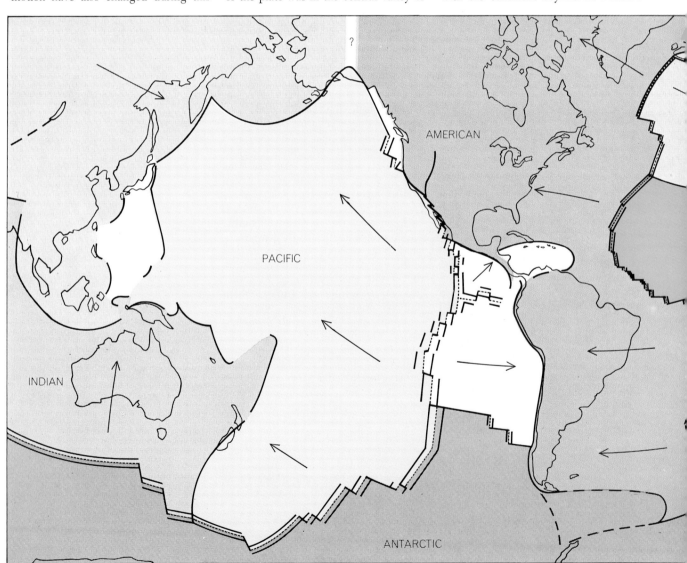

SIX MAJOR PLATES are sufficient to account for the pattern of continental drift inferred to be taking place today. In this model the African plate is assumed to be stationary. Arrows show the direction of motion of the five other large plates, which are generally bounded by ridges or trenches. Several smaller plates, unnamed, also appear. In certain areas, particularly at the junction of

the fragment consists of relatively light rocks it cannot be forced under a continent. The clearest example is the collision of India with what was once the southern margin of Asia. Paleomagnetic work shows that India has been moving northward for the past 100 million years. If it is attached to the plate that is spreading northward and eastward from the Carlsberg Ridge (which runs down the Indian Ocean halfway between Africa and India), then the motion is continuing today. This motion may be the cause of the earthquakes of the Himalayas, and it may also be connected with the formation of the mountains and of the deep sediment-filled trough to the south of them. The exact place where the joint occurs is far from clear and needs study by those with a detailed geological knowledge of northern India.

It seems unlikely that all the continents were collected in a single block for

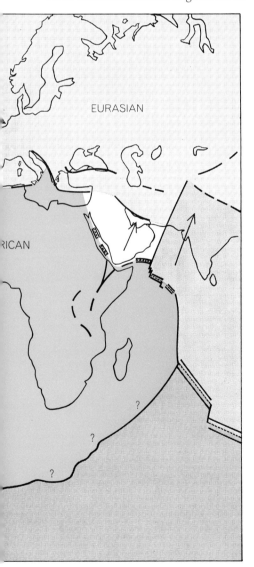

the American and Eurasian plates and in the region south of Africa, it is hard to say just where the boundaries of the plates lie.

4,000 million years and then broke apart and started their wanderings during the past 100 million years. It is more likely that the processes we see today have always been in action and that all through geologic time there have been moving plates carrying continents. We must expect continents to have split many times and formed new oceans and sometimes to have collided and been welded together. We are only at the beginning of the study of pre-Tertiary events; anything that can be said is speculation and is to be taken only as an indication of where to look.

It is virtually certain that the Atlantic did not exist 150 million years ago. Long before that, in the Lower Paleozoic, 650 to 400 million years ago, there was an older ocean in which the sediments now in the Caledonian-Hercynian-Appalachian mountains of Europe and North America were laid down. Perhaps this ocean was closed long before the present Atlantic opened and separated the Appalachian Mountains of eastern North America from their continuation in northwestern Europe.

The Urals, if they are not unique among mountain ranges, are at least exceptional in being situated in the middle of a continent. There is some paleomagnetic evidence that Siberia is a mosaic of fragments that were not originally contiguous; perhaps the Urals were once near the borders of an ocean that divided Siberia from western Russia. Similarly, it is desirable to ask where the ocean was when the Rockies were being formed. A large part of California is moving rapidly northward, and the entire continent has overrun an ocean ridge; clearly the early Tertiary geography must have been very different from that of the present. Such questions are for the future and require that the ideas of moving plates be applied by those with a detailed knowledge of the various areas.

A history of the oceans does not necessarily require an account of the mechanism behind the observed phenomena. Indeed, no very satisfactory account can be given. The traditional view, put forward by Arthur Holmes and Felix A. Vening-Meinesz, supposes that the upper mantle behaves as a liquid when it is subjected to small forces for long periods and that differences in temperature under oceans and continents are sufficient to produce convection cells in the mantle—with rising currents under the mid-ocean ridges and sinking ones under the continents. These hypothetical cells would carry the plates along as on

a conveyor belt and would provide the forces needed to produce the split along the ridge. This view may be correct; it has the advantage that the currents are driven by temperature differences that themselves depend on the position of the continents. Such a back-coupling can produce complicated and varying motions.

On the other hand, the theory is implausible in that convection does not normally happen along lines. It certainly does not happen along lines broken by frequent offsets, as the ridge is. Also it is difficult to see how the theory applies to the plate between the Mid-Atlantic Ridge and the ridge in the Indian Ocean. This plate is growing on both sides, and since there is no intermediate trench the two ridges must be moving apart. It would be odd if the rising convection currents kept exact pace with them. An alternative theory is that the sinking part of the plate, which is denser than the hotter surrounding mantle, pulls the rest of the plate after it. Again it is difficult to see how this applies to the ridge in the South Atlantic, where neither the African nor the American plate has a sinking part.

Another possibility is that the sinking plate cools the neighboring mantle and produces convection currents that move the plates. This last theory is attractive because it gives some hope of explaining the almost enclosed seas, such as the Sea of Japan. These seas have a typical oceanic floor except that the floor is overlain by several kilometers of sediment. Their floors have probably been sinking for long periods. It seems possible that a sinking current of cooled mantle material on the upper side of the plate might be the cause of such deep basins. The enclosed seas are an important feature of the earth's surface and urgently require explanation; in addition to the seas that are developing at present behind island arcs there are a number of older ones of possibly similar origin, such as the Gulf of Mexico, the Black Sea and perhaps the North Sea.

The ideas set out in this attempt at a history of the ocean have developed in the past 10 years. What we have is a sketch of the outlines of a history; a mass of detail needs to be filled in and many major features are quite uncertain. Nonetheless, there is a stage in the development of a theory when it is most attractive to study and easiest to explain, that is while it is still simple and successful and before too many details and difficulties have been uncovered. This is the interesting stage at which plate theory now stands.

III

SOME CONSEQUENCES AND EXAMPLES OF
CONTINENTAL DRIFT

III

SOME CONSEQUENCES
AND EXAMPLES
OF CONTINENTAL DRIFT

INTRODUCTION

Although the five papers in this section are otherwise unrelated, they all show ways in which geological problems can be elucidated through the new ideas of plate tectonics.

The first paper in this section, "The Breakup of Pangaea," was written by Robert S. Dietz and John C. Holden. After reviewing the principles of plate tectonics, Dietz and Holden consider what the global pattern of continents would have been 200 million years ago. They agree with Alfred L. Wegener that they formed a single super-continent, for which Wegener had coined the name Pangaea, and they provide a map of it. Apparently, Pangaea was then on the point of breaking up; by 180 million years ago, it had split apart. At that time, the eastern shore of Pangaea was deeply indented near the equator by a great seaway called the Tethys Sea, which has since closed to form the Himalayan–Alpine mountain chains. It connected with the first opening in the Atlantic (between what is now the Atlantic coast of the United States, on the north, and what is now northwest Africa and South America, on the south) to separate Pangaea into two supercontinents—Laurasia in the Northern Hemisphere and Gondwanaland in the Southern Hemisphere. Gondwanaland also started to break up at about that time. Dietz and Holden illustrate these and later stages (135 and 65 million years ago) by maps. They also provide a map to show what they speculate the world will look like 50 million years from now, if the present directions and rates of motion continue.

Without a doubt, this reconstruction of Pangaea contains errors; indeed, the authors have already changed their minds, and would omit the Kerguelen–Broken Ridge fragment to the west of Australia, because they no longer believe it to be a microcontinent. On the other hand, their sketches show the path that historical geology should follow in the future. Dietz and Holden support the concept that hot spots exist as long-lived, stationary upwellings in the deep mantle, and they use one that is presently beneath the island of Tristan da Cunha as a fixed reference point from which to estimate the past longitudes and latitudes of the continents.

In "Continental Drift and Evolution," Björn Kurtén discusses the evolution of reptiles and mammals in light of the theory of continental drift. Kurtén points out that the age of reptiles lasted 200 million years and produced twenty reptilian orders, whereas the age of mammals lasted only about 65 million years—only about a third as long—but gave rise to thirty mammalian orders. He attributes this to the greater isolation resulting from the separation of continents. It follows that we should expect to find the greatest similarity between the animals of those continents that have separated from one another most recently. Kurtén also shows how examinations of fossils demonstrate that the diversification of many types from common stocks can be traced with some certainty, and goes on to show that the variations in types of living and fossil forms are compatible with the history of drift now generally accepted and outlined in the previous paper. This not only supports the theory of continental drift, but also throws light on the rate and nature of organic evolution itself.

The third paper, "Geosynclines, Mountains and Continent-building" by Robert S. Dietz, was published just as this book was going to press. Consequently, it is discussed in the Conclusion (page 158).

In the fourth paper, "The Afar Triangle," Haroun Tazieff discusses the findings of recent expeditions to one of the world's hottest and most desolate regions. Temperatures reach 134°F in the shade in the Afar Triangle, which is a tangle of deserts, scarps, and salt plains lying in Africa to the west of the junction of the Red Sea and the Gulf of Aden.

Tazieff points out that, among other evidence, the discovery of an early human stone axe encrusted with sea shells shows that this area has been rising rapidly and has but recently emerged from the sea. The uplift has been accompanied by rifting in a northwesterly direction and the extrusion of lavas, chiefly basalts. Tazieff believes that this area is a section of still-spreading ocean floor that has been raised out of the sea so that all the processes are revealed—Iceland is perhaps the only other place yet recognized where this has happened. The area is a small one, on the world scale, but the article combines the excitement of exploration with an unusual illustration of plate tectonics.

The last paper in this anthology is Don L. Anderson's "The San Andreas Fault." Anderson points out that this fault through California forms part of the boundary between the Pacific plate and the North American plate. The former is moving northwestward past the American continents at a rate of $2\frac{1}{2}$ inches per year and is carrying the coastal strip of California to the west of the fault along with it. Although the plates move steadily, the two sides have a tendency to lock together. When this happens, the energy may be stored for many years, until the plates unlock and jump—sometimes as much as several feet—in opposite directions parallel to the fault. This motion produces an earthquake.

This article gives a good account of California earthquakes. It discusses their relative intensity and the scales by which the magnitude and effects of earthquakes are assessed. It discusses the geological history of California in terms of plate tectonics, and it shows that the position of the intersection of the San Andreas fault with the south end of the Sierra Nevada north of Los Angeles is important to interpreting the earthquakes and the geology of California. It concludes that nothing can be done to stop the motion of the plates, and that California will continue to experience earthquakes in the future, just as it has for millions of years. At long intervals, great disasters can be expected, but none of them more cataclysmic than those that have occurred in the past. The prediction of earthquakes is still in its infancy, but Anderson suggests that a dense network of tiltmeters might be of some use. Although earthquakes cannot be stopped, some very recent work has opened the possibility of artificially triggering many small earthquakes, instead of allowing energy to be stored up for release in less frequent but larger shocks. It is possible that water pumped into the ground in the proper places might have the effect of lubricating the fault plane, but little has been learned to show how practical this proposal is.

11

THE BREAKUP OF PANGAEA

ROBERT S. DIETZ AND JOHN C. HOLDEN
October 1970

The history of science is replete with outrageous hypotheses. They are mostly forgotten, as best they should be, but from time to time one of them turns out to be true. So it was with the concept that the earth is a sphere spinning in space, supported by nothing at all. Now it also seems to be with the theory of continental drift, which in its extreme form holds that all the continents were once joined in a single great land mass. Named Pangaea, this universal continent was somehow disrupted, and its fragments—the continents of today—eventually drifted to their present locations.

Over the past three years geologists and geophysicists have been forced to abandon the old dogma that the crust of the earth is essentially fixed and to accept the new heresy that it is quite mobile. The notion that continents can drift thousands of kilometers in a few hundred million years is now generally accepted. Geology therefore finds itself in much the same position that astronomy was in at the time of Copernicus and Galileo. Textbooks are being rewritten to embrace the new mobilistic viewpoint.

Although the theory of continental drift has triumphed, many of its details remain uncertain. Advocates of drift are challenged to say exactly how the present continents fitted together to form Pangaea, or alternatively to reconstruct the two later supercontinents Laurasia and Gondwana, which some theorists prefer to a single all-embracing land mass. The original concept of Pangaea ("all lands") was proposed in the 1920's by Alfred Wegener. Most attempts to improve on his reconstruction have been rather generalized sketches showing how the continents might have been joined. A few workers have made jigsaw

fits with considerable care but without taking advantage of the latest concepts in geotectonics. Recently British theorists have presented detailed reconstructions showing how land masses were juxtaposed before the opening of either the Atlantic or the Indian Ocean, but their solutions show only the relative motions of the masses involved.

In this article we present a reconstruction of Pangaea in which the continents are assembled with cartographic precision. For the first time Pangaea is positioned on the globe in absolute coordinates. This reconstruction is accompanied by four maps that show the breakup and subsequent dispersion of the continents by the end of the four major geologic periods covering the past 180 million years: the Triassic, the Jurassic, the Cretaceous and the Cenozoic.

The guiding rationale for our reconstruction is the drift mechanism associated with plate tectonics and seafloor spreading [*see illustration on page 104*]. According to this concept the earth has a strong lithosphere, or outer shell of rock, about 100 kilometers thick. Presumably in response to forces generated in the asthenosphere, the weak upper mantle of rock underlying the lithosphere, the shell was broken up into a number of separate plates. There are now some 10 major plates, plus numerous additional subplates. The continents resting on these plates were rafted across the surface of the globe.

The mechanism of plate movement is not yet clear. The plates may be pushed, carried by convection cells in the mantle, driven by gravitational forces or pulled. We prefer a model based on pulling; we suspect that plates are colder and heavier at one boundary than elsewhere and thus dive down into the

earth's mantle along "subduction" zones. These zones usually show themselves as deep trenches, which are disposed principally around the periphery of the Pacific. As a result a tear, or rift, widens along the opposite boundary of the plate; this rift is filled by a solid flow of viscous mantle rock and by dikes of molten tholeiitic basalt (a differentiated partial melt of the mantle). Because the mantle rock and its basaltic derivative are both heavier than the granitoid rock of the continents they assume a level about four kilometers below sea level. Consequently such a pulled-apart region always becomes new ocean floor. As two adjacent plates continue to pull apart, basaltic dikes continue to pour into the suboceanic rift, which remains midway between the two plates. This highly symmetrical process, which creates new ocean basins or continuously repaves old ocean floors, is termed seafloor spreading. The rate of spreading, measured from the mid-ocean rift to either plate, is from one centimeter per year (10 kilometers per million years) to several times that figure. This is remarkably rapid by geological standards, being many times faster than mountains are elevated by tectonism or leveled by erosion. For example, the North American plate is moving westward the length of one's body in a lifetime.

The discovery of a mid-ocean ridge system some 40,000 kilometers long, winding through all the ocean basins, was an important prelude to the seafloor-spreading hypothesis. It was soon recognized that the ridge has a fossa, or axial depression, into which dikes of basalt are continuously being injected. This linear depression in the ridge marks the location of the rift. The term "mid-ocean," although appropriate for the part of the ridge system in the Atlantic

SUBCONTINENT OF INDIA, originally attached to what is now Antarctica, made the longest migration of all the drifting land masses: approximately 9,000 kilometers in 200 million years. This picture, taken at an altitude of 650 kilometers from *Gemini XI* in September, 1966, shows all of the subcontinent. The Himalayan mountains, 3,700 kilometers away, are just visible on the horizon.

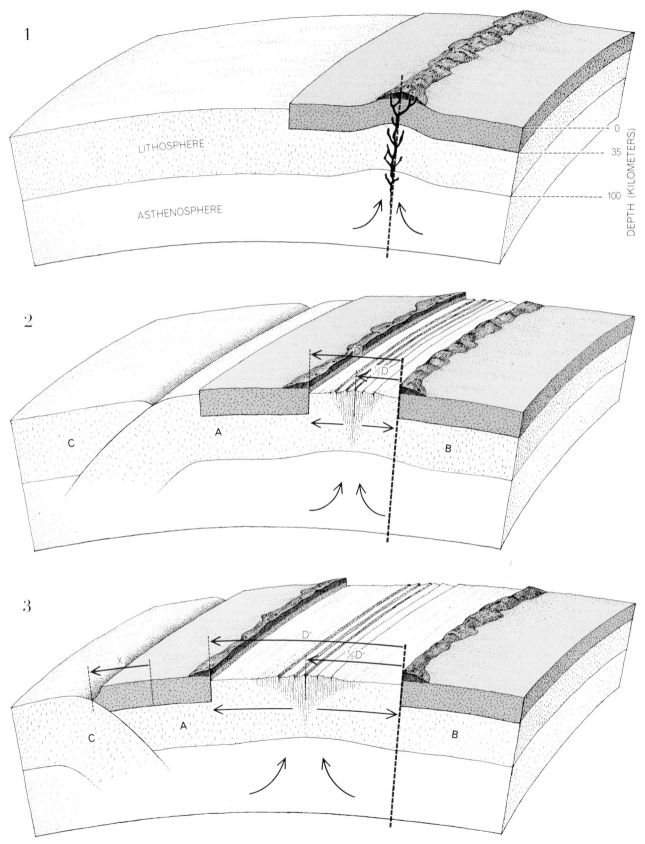

THEORY OF PLATE TECTONICS provides a mechanism for continental drift. The process begins (1) when a spreading rift develops under a continent (*color*) that is resting on a single crustal plate. Molten basalt from the asthenosphere spills out. The second simultaneous requirement for continental drift is the formation of a zone of subduction, or trench, into which oceanic crust of the new moving plate (*A*) is pulled and "consumed" (2). As the new continent carried by plate *A* is rafted to the left, a new ocean basin is created between the two land masses. In the third stage (3) the continent on plate *A* encounters and overrides the trench for some distance (*X*) and eventually reverses, or flips, its direction from west-dipping to east-dipping. Because the continent on plate *B* is here arbitrarily fixed, the mid-ocean rift migrates to the left, remaining in the center of the new ocean basin, whose width is *D'*.

and the Indian Ocean, is a misnomer for the ridge in the Pacific. The Atlantic and the Indian Ocean are rift oceans, formed where continents were once split apart; therefore it is natural for the axis of spreading, marked by the ridge system, to remain in the center of these two oceans. The Pacific, on the other hand, is not a rift ocean; it is clearly the ancestral ocean, and it is becoming smaller as new ocean basins grow. Although the Pacific also has a ridge, it runs north-south well to the east of the ocean's center.

In reality the crustal motions are considerably more complex than the ones we have just outlined. The trenches and rifts apparently migrate, and the opposing plates are also subject to displacements produced by internal shears. The "megashears," the large zones of slippage along plate boundaries, also seem able to accommodate minor amounts of crustal extension or compression. Few of the plates are "ideal" in the sense of being rectilinear, of having a rift matched by an opposing trench and of having these two antithetical zones connected by a megashear. The Antarctic plate, for example, has no trench at all. Perhaps this anomaly is partly explained by the fact that a sphere cannot be covered with rectangles.

We can visualize the continents as being passively rafted over the surface of the globe as embedded plateaus of sialic (granite-like) rock resting on the even larger and thicker crustal plates. The continents have generally maintained their size and shape since the breakup of Pangaea. There have been some accretions with the formation of mountain belts, but these have been mostly confined to the sides of continents facing the Pacific. The sides of continents facing rift oceans (the Atlantic and the Indian Ocean) show little change; hence they can be fitted together almost as neatly as pieces of a jigsaw puzzle.

In contrast, the crustal plates can change in size or shape either by the addition of new ocean floor along the rifts or by the resorption of oceanic crust in trenches. Thus it has been possible for the North American and South American plates moving toward the Pacific to grow larger at first and then smaller as they passed over the great circle of the earth and now converge toward the central Pacific. An even more tortured history is reflected in the complex evolution of the Caribbean Sea region, caught as a "gore" between the North American and South Ameri-

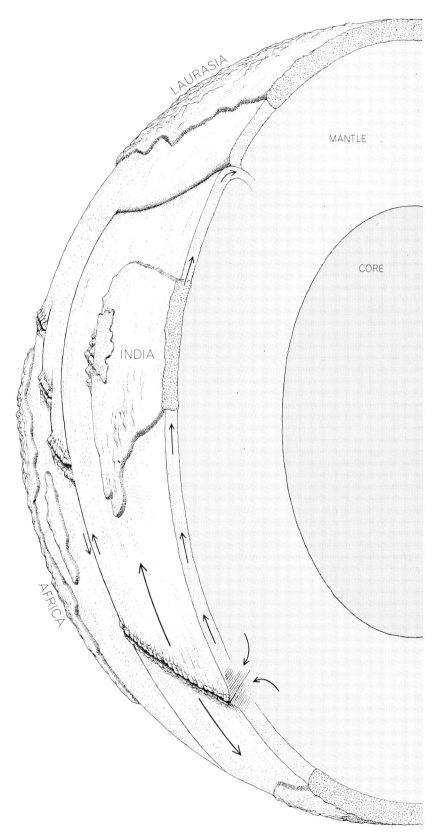

NORTHWARD DRIFT OF INDIA exemplifies how far a land mass can be carried when tectonic conditions are favorable. The plate carrying the Indian land mass is nearly a perfect rectangle, which was sliced away from Antarctica within the primitive universal continent of Pangaea. The plate that rafted India then migrated northward toward and subducted into the Tethyan trench, which ran east-west near the Equator. The plate evidently glided freely along parallel "megashears" on its eastern and western boundaries without interacting with the other crustal plates of the world. India finally collided with and underthrust the southeast margin of Asia, creating the Himalayas, which are thus two plates thick.

can plates, and the Scotia Sea region, similarly trapped between the South American and Antarctica plates. As we shall see, in at least one case two plates evidently collided, producing a mid-continent mountain range: the Himalayas.

In making our reconstruction of Pangaea we selected for fitting not the present coastlines of continents but the contour lines where the continental slope reaches a depth of 1,000 fathoms, or about 2,000 meters [*see illustration below*]. This isobath was selected because it is approximately halfway down the continental slope and thus marks roughly half the height of the vertical walls created when the continents first rifted. On the assumption that these walls subsequently slumped to a condition of stable repose, the 1,000-fathom isobath closely delineates the location of the original break.

For joining the two sides of the Atlantic we have followed, with some modification, the reconstruction proposed by Sir Edward Bullard, J. E. Everett and A. G. Smith of the University of Cambridge. For closing the Indian Ocean we have used the best-fit computer solutions of Walter P. Sproll, a colleague of ours in the Marine Geology and Geophysics Laboratory of the Environmental Science Services Administration. His studies provide precise fits between Australia and Antarctica and between Antarctica and Africa. The three continents together constitute most of Gondwana. Presumably India was also part of the Gondwana complex, but where it was attached remains unclear. Fortunately the pattern of fracture zones in the ocean floor provides crude but useful dead-reckoning tracks showing how the continents drifted. Using such tracks, we have placed the west coast of India against Antarctica rather than against

western Australia, the fit that is often proposed.

Another difficult fit is presented by the bulge of Africa and the bight of North America. The areas of mismatch, particularly that caused by the Florida-Bahamas platform, are sufficiently large for one to reasonably argue that Africa and North America were never joined. On this assumption instead of Pangaea one obtains two unconnected supercontinents as the antecedent land masses: Laurasia in the Northern Hemisphere and Gondwana in the Southern. This version of the continental-drift theory has important adherents.

We nevertheless prefer the Pangaea reconstruction; in our view the areas of mismatch can reasonably be regarded as modifications that arose after Africa and North America began drifting apart. We regard the Florida-Bahamas platform as a sedimentary infilling of a small ocean basin that appeared when Africa

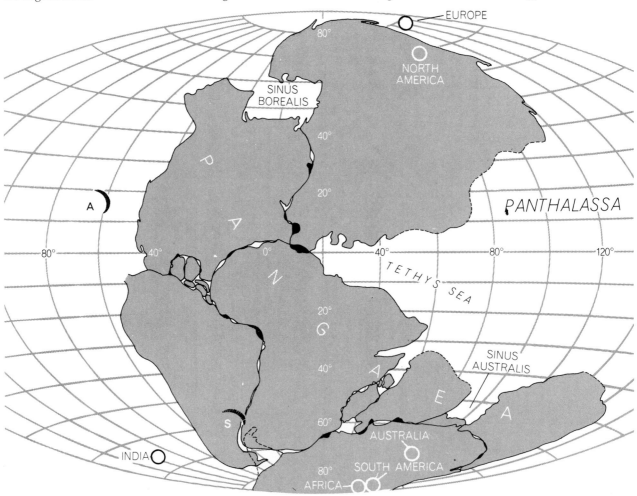

UNIVERSAL LAND MASS PANGAEA may have looked like this 200 million years ago. Panthalassa was the ancestral Pacific Ocean. The Tethys Sea (the ancestral Mediterranean) formed a large bay separating Africa and Eurasia. The relative positions of the continents, except for India, are based on best fits made by computer, using the 1,000-fathom isobath to define continental boundaries.

When the continents are arranged as shown, the relative locations of the magnetic poles in Permian times are displaced to the positions marked by circles. Ideally these positions should cluster near the geographic poles. The hatched crescents (*A and S*) serve as modern geographic reference points; they represent the Antilles arc in the West Indies and Scotia arc in the extreme South Atlantic.

and North America first began to pull apart. Without this assumption the platform unaccountably overlaps a large portion of the bulge of Africa [*see illustration on page 112*].

According to our reconstruction, Pangaea was a land mass of irregular outline surrounded by the universal ocean of Panthalassa: the ancestral Pacific. The fit between North America and Africa provides the principal connection between the future block of northern continents and the future group of southern ones. On the east the Tethys Sea, a large triangular bight, separated Eurasia from Africa; the present Mediterranean Sea is a remnant of the Tethys. Other major indentations in the outline of Pangaea (adapting terminology from the moon) can be named Sinus Borealis, the ancestral Arctic Ocean, and Sinus Australis, a southern bay off the Tethys separating India from Australia. Our fully closed reconstruction of the Central American

region is problematical. An alternate possibility is that the Gulf of Mexico is the remnant of an oceanic arm extending into the Americas from Panthalassa— a Sinus Occidentalis.

When measured down to the 1,000-fathom isobath, the total area of Pangaea was 200,000 square kilometers, or 40 percent of the earth's surface—equal to the area of the present continents measured to the same isobath. When the future continents were still part of Pangaea, they were generally to the south and east of their present location, so that the amount of land in the two hemispheres was almost equally balanced. (Today two-thirds of all the land lies north of the Equator.) The Y-shaped junction connecting North America, South America and Africa was located in the South Atlantic not far from the present position of Ascension Island. If New York had been in existence at the time, it would have been on the Equator

and at longitude 10 degrees east (rather than 74 degrees west). Spain would also have been on the Equator, but it would have been near its present longitude. Japan would have been in the Arctic, well north of its position today. India and Australia would have bordered the Antarctic, far to the south of where they are now.

The great event that broke up Pangaea and set its fragments adrift evidently began no more than 200 million years ago, or in the last few percent of geologic time. There may have been— indeed, there probably was—"predrift drift" that assembled Pangaea from two or more smaller land masses. The evidence is still scanty, however, and does not bear directly on this discussion.

We take the immediate prelude to the breakup of Pangaea to be the first large outpourings of basaltic rock along the continental margins being es-

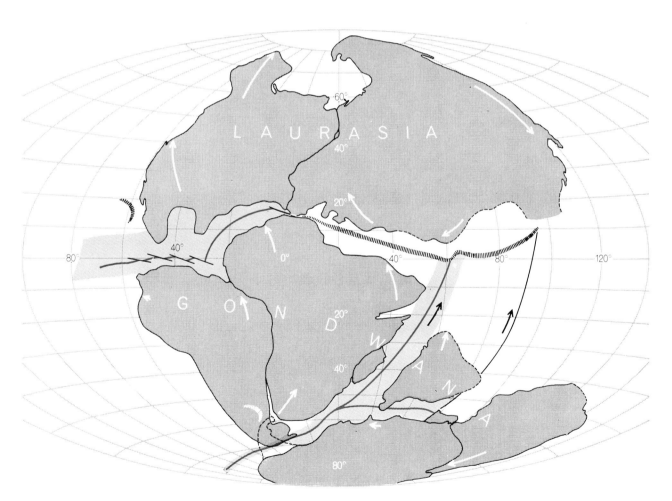

AFTER 20 MILLION YEARS OF DRIFT, at the end of the Triassic period 180 million years ago, the northern group of continents, known as Laurasia, has split away from the southern group, known as Gondwana. The latter has started to break up: India has been set free by a Y-shaped rift (*heavy line in color*), which has also begun to isolate the Africa–South America land mass from Antarctica-Australia. The Tethyan trench (*hatched lines in black*), a zone of crustal uptake, runs from Gibralter to the general area of Borneo. Black lines and black arrows denote megashears, zones of slippage along plate boundaries. The white arrows indicate the vector motions of the continents since drift began. Oceanic areas tinted in color represent new ocean floor created by sea-floor spreading.

tablished by rifting. The Triassic New-ark series of basaltic flows along the east coast of the U.S. is a good example. Measurements of radioactivity indicate that the most ancient of these rocks are about 200 million years old, yielding a date that coincides with the middle of the Triassic period. As we interpret the evidence, two extensive rifts were initiated in Pangaea about 200 million years ago, which resulted in the opening of the Atlantic and the Indian Ocean by the end of the Triassic period 180 million years ago [see illustration on preceding page]. The northern rift split Pangaea from east to west along a line slightly to the north of the Equator and created Laurasia, composed of North America and Eurasia. The Laurasian land mass evidently rotated clockwise as a single plate around a pole of rotation that is now in Spain, creating a western "Mediterranean" that ultimately became part of the Gulf of Mexico and the Caribbean Sea. The southern rift split South America and Africa as a

single land mass away from the remainder of Gondwana, consisting of Antarctica, Australia and India. Soon afterward (if not simultaneously) India was severed from Antarctica by a smaller rift to begin its rapid drift northward.

During the Jurassic period, from 180 to 135 million years ago, the direction of drift established by the Triassic rifts continued, further opening up the Atlantic and the Indian Ocean [see illustration below]. As North America drifted to the northwest, the Atlantic became more than 1,000 kilometers wide and probably remained fully connected to the Pacific. The east coast of the present U.S. ran almost east and west at a latitude of about 25 degrees north, so that coral reefs were able to grow all along the edge of the Atlantic continental shelf to the present Grand Banks, off Nova Scotia.

During the 45-million-year Jurassic period the Atlantic rift extended northward, blocking out the Labrador coastline and possibly initiating the opening

up of the Labrador Sea between North America and Greenland. The interaction between the African and Eurasian plates forced the region of Spain to rotate counterclockwise 35 degrees, opening up the Bay of Biscay. The Tethys Sea, forerunner of the Mediterranean, continued to close at its eastern end. The Tethys was not only a zone of crustal subduction, or trench, but also a zone of shear along which Eurasia slid westward with respect to Africa. The compression associated with the Tethys trench raised bordering mountains composed of deep-water sediments.

At the close of the Jurassic an incipient rift began splitting South America away from Africa, entering from the south and working only as far north as where Nigeria is today. The tectonic situation first resembled the one now found in the rift zone along the backbone of high Africa (the region from Ethiopia to Tanzania) and then gradually opened farther to form a body of water resembling the Red Sea of today.

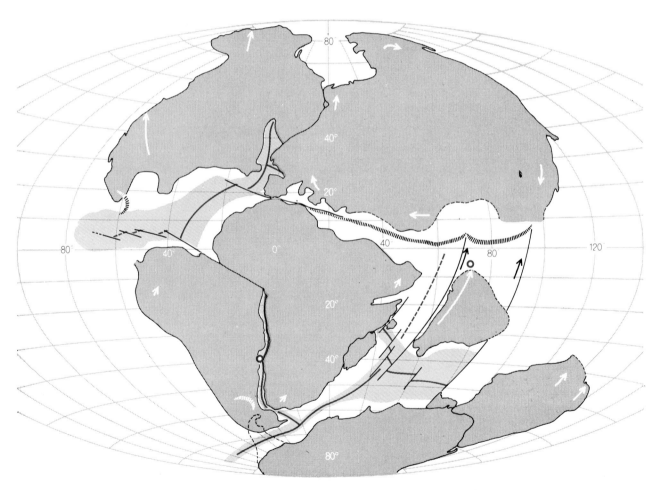

AFTER 65 MILLION YEARS OF DRIFT, at the end of the Jurassic period 135 million years ago, the North Atlantic and the Indian Ocean have opened considerably. The birth of the South Atlantic has been initiated by a rift. The rotation of the Eurasian land mass has begun to close the eastern end of the Tethys Sea. The Indian plate is about to pass over a thermal center (colored dot) that will soon pour out basalt to form the Deccan plateau. Later the hot spot will create the Chagos-Laccadive ridge in the Indian Ocean. Similarly, in the South Atlantic the Walvis thermal center (colored dot) will create the Walvis and Rio Grande "thread ridges."

At first freshwater sediments created thick deposits in pockets opened by faults; these sediments were overlain by deposits of salt.

By the end of the Cretaceous period, some 70 million years later (and 65 million years ago), the rupture of South America and Africa was complete, and the South Atlantic had widened rapidly to at least 3,000 kilometers [*see illustration below*]. Meanwhile the rift in the North Atlantic had switched from the west side of Greenland to the east side, blocking out its eastern margin (without, however, penetrating to the Arctic Ocean). Africa had drifted northward about 10 degrees and continued its counterclockwise rotation as the Eurasian plate rotated slowly clockwise. These two opposed motions nearly closed the eastern end of the Tethys Sea. The slow westward rotation of Antarctica continued. All the continents were now blocked out except for the remaining connection between Greenland and

northern Europe and between Australia and Antarctica.

Although it is not shown on our maps, an extensive north-south trench system must have existed in the ancient Pacific to consume by subduction the rapid westward drift of the two plates carrying North and South America. North America presumably encountered this trench in the late Jurassic and early Cretaceous, with the result that the Franciscan fold belt, the predecessor of the California Coast Ranges, was accreted to the western margin of the continent. It appears that the trench was eventually overridden and "stifled" by North America's continued westward drift. Such trenches have the capacity to resorb ocean crust but not the lighter granitic crust of continents.

At about the same time, or soon afterward, South America first encountered the Andean trench and began to displace the trench westward, without ever overriding it. The early Andean fold belt resulted from this encounter. It seems

likely that the trench originally dipped toward the west but was flipped over to its present eastward dip.

In the Cenozoic period (from 65 million years ago to the present) the continents drifted to the positions we observe today. The mid-Atlantic rift propagated into the Arctic basin, finally detaching Greenland from Europe [*see top illustration on next two pages*]. There were three other major developments during the Cretaceous: (1) the two Americas were rejoined by the Isthmus of Panama, created by volcanism and the arching upward of the earth's mantle, (2) the Indian land mass completed its remarkable journey northward by colliding with the underbelly of Asia and (3) Australia was rifted away from Antarctica and drifted northward to its present position.

In the collision of India with Asia the northern margin of the Indian plate was subducted below the Asiatic plate, creating the Himalayas. On India's passage to the north early in the Cenozoic its western margin crossed a fixed source of

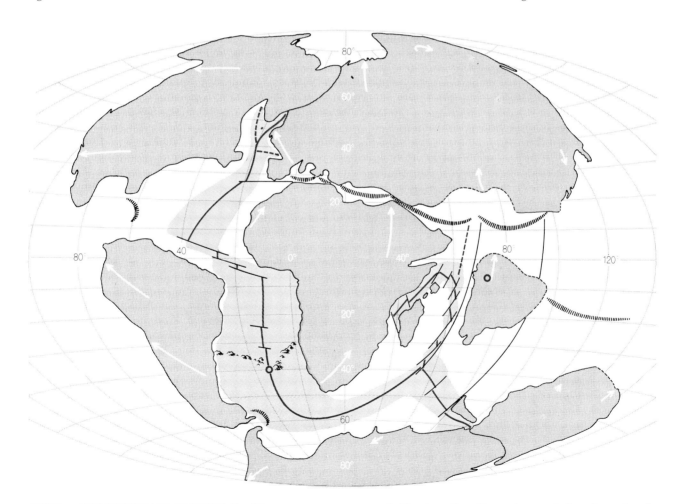

AFTER 135 MILLION YEARS OF DRIFT, 65 million years ago at the end of the Cretaceous period, the South Atlantic has widened into a major ocean. A new rift has carved Madagascar away from Africa. The rift in the North Atlantic has switched from the west side to the east side of Greenland. The Mediterranean Sea is clearly recognizable. Australia still remains attached to Antarctica. An extensive north-south trench (*not shown*) must also have existed in the Pacific to absorb the westward drift of the North American and South American plates. Note that the central meridian in all these reconstructions is 20 degrees east of the Greenwich meredian.

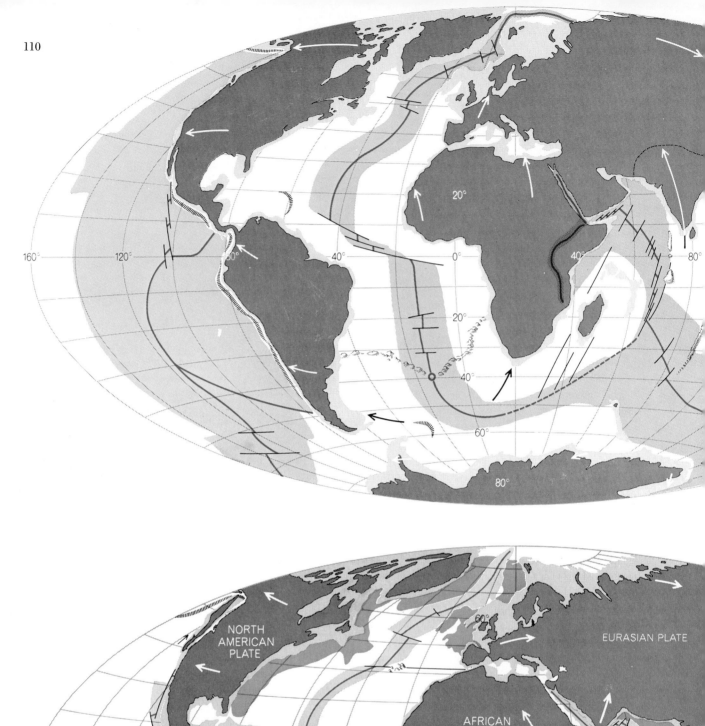

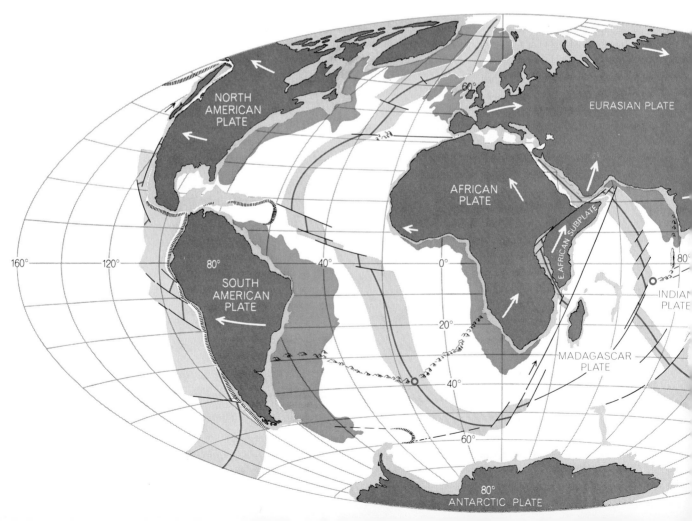

WORLD AS IT LOOKS TODAY was produced in the past 65 million years in the Cenozoic period. Nearly half of the ocean floor was created in this geologically brief period, as shown by the areas stippled in color. India completed its flight northward by colliding with Asia and a rift has separated Australia from Antarctica. The North Atlantic rift finally entered the Arctic Ocean, fissioning Laurasia. The widening gap between South America and Africa is closely traced by the thread ridges produced by the Walvis thermal center. The Antilles and Scotia arcs now occupy their proper positions with respect to neighboring land masses.

basaltic magma rising from the earth's upper mantle near the Equator. Molten rock erupted through the crust and poured onto the Indian subcontinent, laying down the basalts of the Deccan plateau. Even after India had left the hot spot behind, magma continued to stream out on the ocean floor, producing the Chagos-Laccadive ridge, which became covered with coral as it subsided into the Indian Ocean. Finally, a branch of the Indian Ocean rift split Arabia away from Africa, creating the Gulf of Aden and the Red Sea, and a spur of this rift meandered west and south into Africa.

Less pronounced changes during the Cenozoic period included the partial closing of the Caribbean region and the continued widening of the South Atlantic as new ocean crust was emplaced by sea-floor spreading. As the Atlantic continued to open in the far north the northwestward movement of the Eurasian land mass was halted and reversed, simultaneously reversing its sense of slippage with respect to Africa. The new direction of shear has been strongly impressed on the tectonic character of the Mediterranean and the Near East. The major north-south rift in the Indian Ocean largely ceased spreading and became instead a megashear that accommodated the counterclockwise and northward rotation of the African plate.

The reader will have observed that our maps of continental drift show more than relative positions and motions; the land masses, beginning with Pangaea itself, are assigned absolute geographic coordinates. Since this has not been attempted before we shall briefly describe how we arrived at our results. In the mobile world of plate tectonics one must assume that all parts of the crust are capable of moving and almost surely have moved.

After an extensive search for some absolute reference point, we finally concluded that the Walvis thermal center, or hot spot, might provide what we sought. In reaching this conclusion we accepted a hypothesis put forward by J. Tuzo Wilson of the University of Toronto. He had suggested that the Walvis ridge and the Rio Grande ridge in the South Atlantic are nemataths, or "thread ridges" of basalt, that had been poured onto the spreading ocean floor from a fixed lava orifice rising from a deep, stagnant region of the mantle. As new floor was carried past the orifice, lava would periodically pour out and form a small volcanic cone. By observing the location of succeeding cones as they merged into a ridge one can establish the absolute direction taken by the crust in that region. A study of the Walvis and Rio Grande ridges enabled us to establish not only the drift of the South American plate with respect to the African plate but also any motion the two plates may have had in some other direction [*see illustration on page 113*].

Unfortunately the Walvis hot spot did not exist earlier than about 140 million years ago, so that its usefulness as a fixed point does not go back earlier than the end of the Jurassic period. To trace crustal motions during the first 60 million years after the breakup of Pangaea one has to rely on dead reckoning. We have made the assumption that Antarctica has moved very little from its original location when it was part of Pan-

WORLD 50 MILLION YEARS FROM NOW may look something like this. The authors have extrapolated present-day plate movements to indicate how the continents will have drifted by the end of what they propose to call the Psychozoic era (the age of awareness). The Antarctic remains essentially fixed but may rotate slightly clockwise. The Atlantic (particularly the South Atlantic) and the Indian Ocean continue to grow at the expense of the Pacific. Australia drifts northward and begins rubbing against the Eurasian plate. The eastern portion of Africa is split off, while its northward drift closes the Bay of Biscay and virtually collapses the Mediterranean. New land area is created in the Caribbean by compressional uplift. Baja California and a sliver of California west of the San Andreas fault are severed from North America and begin drifting to the northwest. In about 10 million years Los Angeles will be abreast of San Francisco, still fixed to the mainland. In about 60 million years Los Angeles will start sliding into the Aleutian trench.

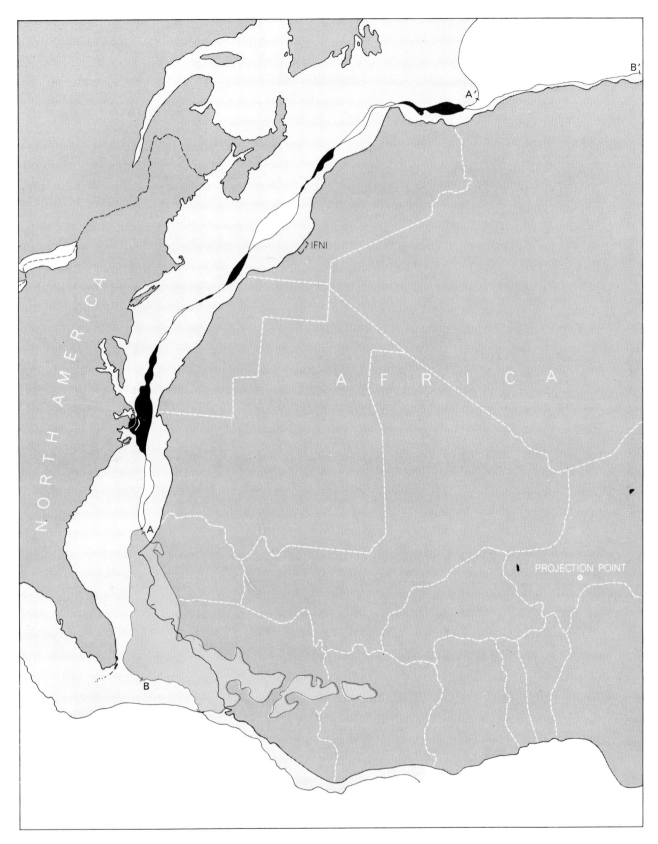

FIT OF AFRICA AGAINST NORTH AMERICA was made by the authors' colleague Walter P. Sproll with the aid of a computer. As in the reconstruction of Pangaea, it is assumed that each continent actually extends out into the ocean and halfway down the continental slope, where the ocean reaches a depth of 1,000 fathoms. The North American "coast" between A and A' was matched for best fit to the African "coast" between B and B'. White areas are gaps in the fit; black areas are overlaps. The overlap produced by the Bahamas platform, an enormous area half the size of Texas, is specially depicted in dark color. The authors propose that the platform represents an accumulation of sediments followed by coral growth after the two continents became separated. The largest gap in the proposed fit between the two continents is found off the Spanish enclave of Ifni. The Ifni gap may have been created when a small section of Africa split off and was translated 190 kilometers to the southwest, forming the eastern group of the Canary Islands.

gaea. This seems reasonable because the Antarctica plate is entirely surrounded by a system of rifts and megashears; there is no associated trench toward which the plate would tend to move away from its polar position.

Independent support for this assumption is obtained by plotting the position of the North and South poles before the dispersal of Pangaea. These positions are obtained by studying the direction of magnetization in rocks of the Permian period, as obtained by E. Irving of the Dominion Observatory in Canada and by other workers. We plotted the Permian pole positions with respect to each continent as it exists today and then rotated these pole positions as needed to assemble the continents into our version of Pangaea. By this method the pole positions should ideally cluster at one of the geographic poles. Actually there is some scatter, as can be seen in our reconstruction of Pangaea [see illustration on page 106], but all the positions do fall within either the Arctic Circle or the Antarctic Circle.

We can now summarize how the continents have moved in time and space. The two Americas have drifted a long way, generally westward. North America has drifted more than 8,000 kilometers west northwest; the tip of Florida once lay in the South Atlantic near the present position of Ascension Island. Moving toward the Tethyan trench system, India and Australia were carried far to the north. Africa rotated counterclockwise perhaps 20 degrees as the Eurasian land mass, similarly moving toward the Tethyan trench, rotated clockwise a roughly equal amount. India's remarkable flight is probably attributable to its being rafted on an "ideal" plate. Approximately rectangular, the Indian plate was sliced away from Pangaea by a rift along what is now India's east coast and then was free to move northward toward a major trench. This northward movement was facilitated by two parallel megashears.

Decades ago Wegener proposed that the drift of the continents was vectored by forces he termed *Westwanderung* (westward drift) and *Polarfluchtkraft* (flight from the poles). Although real, these forces are minuscule and not likely to be the underlying cause of drift. Our solution, however, does support Wegener's hypothesis of a westward flight, which, like the slip of the atmosphere, directly opposes the earth's rotation. We have also inferred a latitudinal drift, but from the South Pole only, or, paraphrasing Wegener's terminology, a *Sudpolarfluchtkraft.*

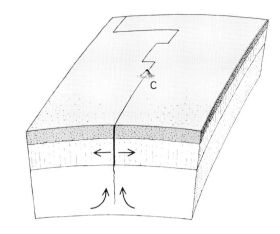

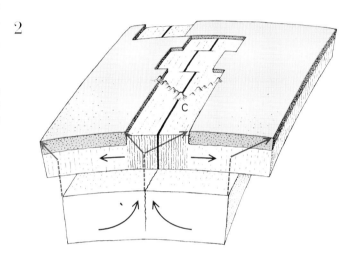

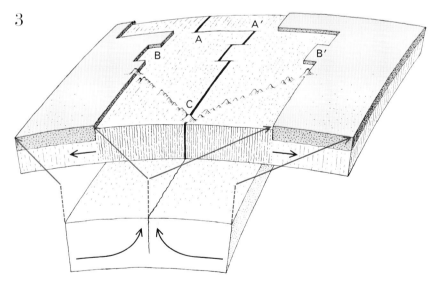

SEPARATION OF SOUTH AMERICAN AND AFRICAN PLATES can be traced in absolute geographic coordinates by observing the orientation of the thread ridges, a V-shaped stream of volcanoes, produced by the Walvis thermal center (C). The hot spot has evidently been pouring out magma from a source deep in the mantle for the past 140 million years. The three-part diagram illustrates a hypothesis first proposed by J. Tuzo Wilson of the University of Toronto. The thread ridges show that the South American and African plates have been not only drifting rapidly apart but also migrating northward. Features such as the strike of the ridge-ridge transform faults (A–A') and matching indentations on opposing continents (B–B') can do more than indicate the relative motion of two plates.

12

CONTINENTAL DRIFT AND EVOLUTION

BJÖRN KURTÉN

March 1969

The history of life on the earth, as it is revealed in the fossil record, is characterized by intervals in which organisms of one type multiplied and diversified with extraordinary exuberance. One such interval is the age of reptiles, which lasted 200 million years and gave rise to some 20 reptilian orders, or major groups of reptiles. The age of reptiles was followed by our own age of mammals, which has lasted for 65 million years and has given rise to some 30 mammalian orders.

The difference between the number of reptilian orders and the number of mammalian ones is intriguing. How is it that the mammals diversified into half again as many orders as the reptiles in a third of the time? The answer may lie in the concept of continental drift, which has recently attracted so much attention from geologists and geophysicists [see the article "The Confirmation of Continental Drift," by Patrick M. Hurley, beginning on page 57]. It now seems that for most of the age of reptiles the continents were collected in two supercontinents, one in the Northern Hemisphere and one in the Southern. Early in the age of mammals the two supercontinents had apparently broken up into the continents of today but the present connections between some of the continents had not yet formed. Clearly such events would have had a profound effect on the evolution of living organisms.

The world of living organisms is a world of specialists. Each animal or plant has its special ecological role. Among the mammals of North America, for instance, there are grass-eating prairie animals such as the pronghorn antelope, browsing woodland animals such as the deer, flesh-eating animals specializing in large game, such as the mountain lion, or in small game, such as the fox, and so on. Each order of mammals comprises a number of species related to one another by common descent, sharing the same broad kind of specialization and having a certain physical resemblance to one another. The order Carnivora, for example, consists of a number of related forms (weasels, bears, dogs, cats, hyenas and so on), most of which are flesh-eaters. There are a few exceptions (the aardwolf is an insect-eating hyena and the giant panda lives on bamboo shoots), but these are recognized as late specializations.

Radiation and Convergence

In spite of being highly diverse, all the orders of mammals have a common origin. They arose from a single ancestral species that lived at some unknown time in the Mesozoic era, which is roughly synonymous with the age of reptiles. The American paleontologist Henry Fairfield Osborn named the evolution of such a diversified host from a single ancestral type "adaptive radiation." By adapting to different ways of life—walking, climbing, swimming, flying, plant-eating, flesh-eating and so on—the descendant forms come to diverge more and more from one another. Adaptive radiation is not restricted to mammals; in fact we can trace the process within every major division of the plant and animal kingdoms.

The opposite phenomenon, in which stocks that were originally very different gradually come to resemble one another through adaptation to the same kind of life, is termed convergence. This too seems to be quite common among mammals. There is a tendency to duplication—indeed multiplication—of orders performing the same function. Perhaps the most remarkable instance is found among the mammals that have specialized in large-scale predation on termites and ants in the Tropics. This ecological niche is filled in South America by the ant bear *Myrmecophaga* and its related forms, all belonging to the order Edentata. In Asia and Africa the same role is played by mammals of the order Pholidota: the pangolins, or scaly anteaters. In Africa a third order has established itself in this business: the Tubulidentata, or aardvarks. Finally, in Australia there is the spiny anteater, which is in the order Monotremata. Thus we have members of four different orders living the same kind of life.

One can cite many other examples. There are, for instance, several living and extinct orders of hoofed herbivores. There are two living orders (the Rodentia, or rodents, and the Lagomorpha, or rabbits and hares) whose chisel-like incisor teeth are specialized for gnawing. Some extinct orders specialized in the same way, and an early primate, an ice-age ungulate and a living marsupial have also intruded into the "rodent niche" [*see top illustration on page 119*]. This kind of duplication, or near-duplication, is an essential ingredient in the richness of the mammalian life that unfolded during the Cenozoic era, or the age of mammals. Of the 30 or so orders of land-dwelling mammals that appeared during this period almost two-thirds are still extant.

The Reptiles of the Cretaceous

The 65 million years of the Cenozoic are divided into two periods: the long Tertiary and the brief Quaternary, which includes the present [*see illustration on page 116*]. The 200-million-year age of reptiles embraces the three periods of the Mesozoic era (the Triassic, the Jurassic and the Cretaceous) and the final period (the Permian) of the preceding era. It is instructive to compare the number

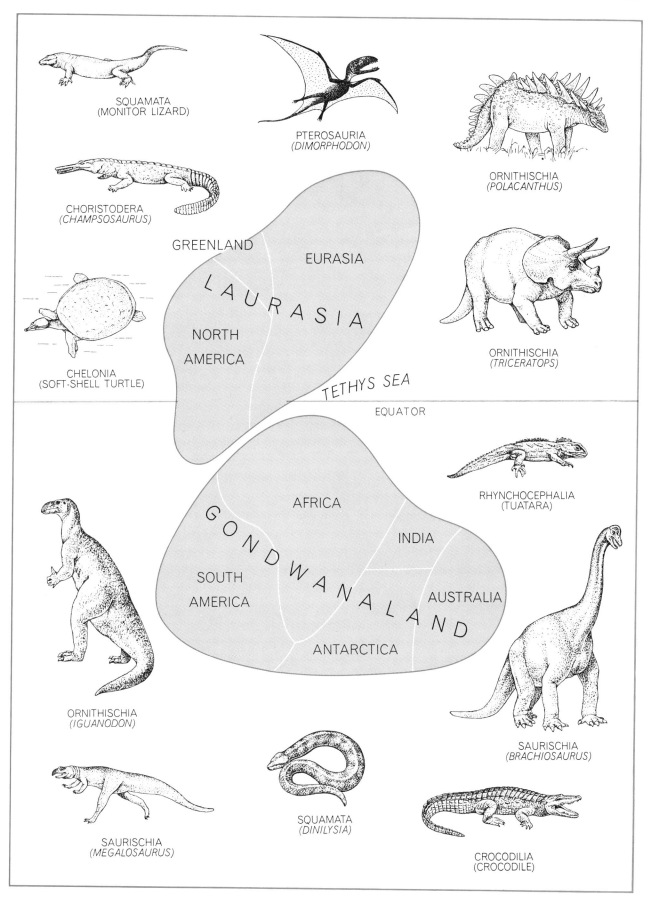

TWO SUPERCONTINENTS of the Mesozoic era were Laurasia in the north and Gondwanaland in the south. The 12 major types of reptiles, represented by typical species, are those whose fossil remains are found in Cretaceous formations. Most of the orders inhabited both supercontinents; migrations were probably by way of a land bridge in the west, where the Tethys Sea was narrowest.

116

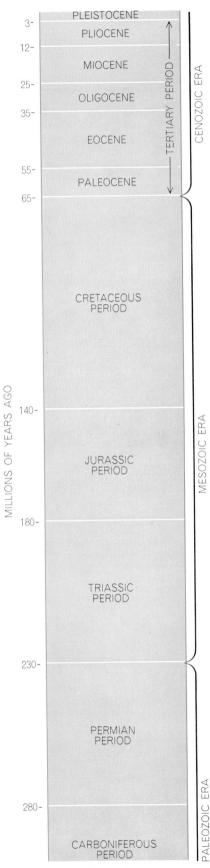

SIX PERIODS of earth history were occupied by the age of reptiles and the age of mammals. The reptiles' rise began 280 million years ago, in the final period of the Paleozoic era. Mammals replaced reptiles as dominant land animals 65 million years ago.

of reptilian orders that flourished during some Mesozoic interval about as long as the Cenozoic era with the number of mammalian orders in the Cenozoic. The Cretaceous period is a good candidate. Some 75 million years in duration, it is only slightly longer than the age of mammals. Moreover, the Cretaceous was the culmination of reptilian life and its fossil record on most continents is good. In the Cretaceous the following orders of land reptiles were extant:

Order Crocodilia: crocodiles, alligators and the like. Their ecological role was amphibious predation; their size, medium to large.

Order Saurischia: saurischian dinosaurs. These were of two basic types: bipedal upland predators (Theropoda) and very large amphibious herbivores (Sauropoda).

Order Ornithischia: ornithischian dinosaurs. Here there were three basic types: bipedal herbivores (Ornithopoda), heavily armored quadrupedal herbivores (Stegosauria and Ankylosauria) and horned herbivores (Ceratopsia).

Order Pterosauria: flying reptiles.

Order Chelonia: turtles and tortoises.

Order Squamata: The two basic types were lizards (Lacertilia) and snakes (Serpentes). Both had the same principal ecological role: small to medium-sized predator.

Order Choristodera (or suborder in the order Eosuchia): champsosaurs. These were amphibious predators.

One or two other reptilian orders may be represented by rare forms. Even if we include them, we get only eight or nine orders of land reptiles in Cretaceous times. One could maintain that an order of reptiles ranks somewhat higher than an order of mammals; some reptilian orders include two or even three basic adaptive types. Even if these types are kept separate, however, the total rises only to 12 or 13. Furthermore, there seems to be only one clear-cut case of ecological duplication: both the crocodilians and the champsosaurs are sizable amphibious predators. (The turtles cannot be considered duplicates of the armored dinosaurs. For one thing, they were very much smaller.) A total of somewhere between seven and 13 orders over a period of 75 million years seems a sluggish record compared with the mammalian achievement of perhaps 30 orders in 65 million years. What light can paleogeography shed on this matter?

The Mesozoic Continents

The two supercontinents of the age of reptiles have been named Laurasia (after

Laurentian and Eurasia) and Gondwanaland (after a characteristic geological formation, the Gondwana). Between them lay the Tethys Sea (named for the wife of Oceanus in Greek myth, who was mother of the seas). Laurasia, the northern supercontinent, consisted of what would later be North America, Greenland and Eurasia north of the Alps and the Himalayas. Gondwanaland, the southern one, consisted of the future South America, Africa, India, Australia and Antarctica. The supercontinents may have begun to split up as early as the Triassic period, but the rifts between them did not become effective barriers to the movement of land animals until well into the Cretaceous, when the age of reptiles was nearing its end.

When the mammals began to diversify in the late Cretaceous and early Tertiary, the separation of the continents appears to have been at an extreme. The old ties were sundered and no new ones had formed. The land areas were further fragmented by a high sea level; the waters flooded the continental margins and formed great inland seas, some of which completely partitioned the continents. For example, South America was cut in two by water in the region that later became the Amazon basin, and Eurasia was split by the joining of the Tethys Sea and the Arctic Ocean. In these circumstances each chip of former supercontinent became the nucleus for an adaptive radiation of its own, each fostering a local version of a balanced fauna. There were at least eight such nuclei at the beginning of the age of mammals. Obviously such a situation is quite different from the one in the age of reptiles, when there were only two separate land masses.

Where the Reptiles Originated

The fossil record contains certain clues to some of the reptilian orders' probable areas of origin. The immense distance in time and the utterly different geography, however, make definite inferences hazardous. Let us see what can be said about the orders of Cretaceous reptiles (most of which, of course, arose long before the Cretaceous):

Crocodilia. The earliest fossil crocodilians appear in Middle Triassic formations in a Gondwanaland continent (South America). The first crocodilians in Laurasia are found in Upper Triassic formations. Thus a Gondwanaland origin is suggested.

Saurischia. The first of these dinosaurs appear on both supercontinents in the Middle Triassic, but they are more

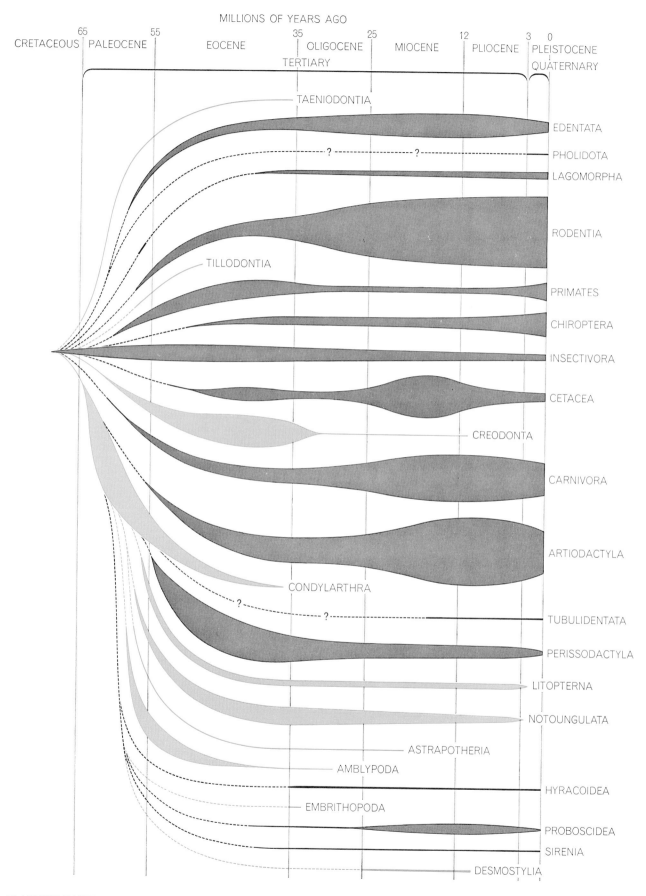

ADAPTIVE RADIATION of the mammals has been traced from its starting point late in the Mesozoic era by Alfred S. Romer of Harvard University. Records for 25 extinct and extant orders of placental mammals are shown here. The lines increase and decrease in width in proportion to the abundance of each order. Extinct orders are shown in color; broken lines mean that no fossil record exists during the indicated interval and question marks imply doubt about the suggested ancestral relation between some orders.

varied in the south. A Gondwanaland origin is very tentatively suggested.

Ornithischia. These dinosaurs appear in the Upper Triassic of South Africa (Gondwanaland) and invade Laurasia somewhat later. A Gondwanaland origin is indicated.

Pterosauria. The oldest fossils of flying reptiles come from the early Jurassic of Europe. They represent highly specialized forms, however, and their antecedents are unknown. No conclusion seems possible.

Chelonia. Turtles are found in Triassic formations in Laurasia. None are found in Gondwanaland before Cretaceous times. This suggests a Laurasian origin. On the other hand, a possible forerunner of turtles appears in the Permian of South Africa. If the Permian form was in fact ancestral, a Gondwanaland origin would be indicated. In any case, the order's main center of evolution certainly lay in the northern supercontinent.

Squamata. Early lizards are found in the late Triassic of the north, which may suggest a Laurasian origin. Unfortunately the lizards in question are aberrant gliding animals. They must have had a long history, of which we know nothing at present.

Choristodera. The crocodile-like champsosaurs are found only in North America and Europe, and so presumably originated in Laurasia.

The indications are, then, that three orders of reptiles—the crocodilians and the two orders of dinosaurs—may have originated in Gondwanaland. Three others—the turtles, the lizards and snakes and the champsosaurs—may have originated in Laurasia. The total number of basic adaptive types in the Gondwanaland group is six; the Laurasia group has four. The Gondwanaland radiation may well have been slightly richer than the Laurasian because it seems that the southern supercontinent was somewhat larger and had a slightly more varied climate. Laurasian climates seem to have been tropical to temperate. Southern parts of Gondwanaland were heavily glaciated late in the era preceding the Mesozoic, and its northern shores (facing the Tethys Sea) had a fully tropical climate.

Although some groups of reptiles, such as the champsosaurs, were confined to one or another of the supercontinents, most of the reptilian orders sooner or later spread into both of them. This means that there must have been ways for land animals to cross the Tethys Sea. The Tethys was narrow in the west and wide to the east. Presumably whatever land connection there was—a true land bridge or island stepping-stones—was located in the western part of the sea. In any case, migration along such routes meant that there was little local differentiation among the reptiles of the Mesozoic era. It was over an essentially uniform reptilian world that the sun finally set at the end of the age of reptiles.

Early Mammals of Laurasia

The conditions of mammalian evolution were radically different. In early and middle Cretaceous times the connections between continents were evidently close enough for primitive mammals to spread into all corners of the habitable world. As the continents drifted farther apart, however, populations of these primitive forms were gradually isolated from one another. This was particularly the case, as we shall see, with the mammals that inhabited the daughter continents of Gondwanaland. Among the Laurasian continents North America was drifting away from Europe, but at the beginning of the age of mammals the distance was not great and there is good evidence that some land connection remained well into the early Tertiary. North American and European mammals were practically identical as late as early Eocene times. Furthermore, throughout the Cenozoic era there was a connection between Alaska and Siberia, at least intermittently, across the Bering Strait. On the other hand, the inland sea extending from the Tethys to the Arctic Ocean formed a complete barrier to direct migration between Europe and Asia in the early Tertiary. Migrations could take place only by way of North America.

In this way the three daughter continents of ancient Laurasia formed three semi-isolated nuclear areas. Many orders of mammals arose in these Laurasian nuclei, among them seven orders that are now extinct but that covered a wide spectrum of specialized types, including primitive hoofed herbivores, carnivores, insectivores and gnawers. The orders of mammals that seem to have arisen in the northern daughter continents and that are extant today are:

Insectivora: moles, hedgehogs, shrews and the like. The earliest fossil insectivores are found in the late Cretaceous of North America and Asia.

Chiroptera: bats. The earliest-known bat comes from the early Eocene of North America. At a slightly later date bats were also common in Europe.

Primates: prosimians (for example, tarsiers and lemurs), monkeys, apes, man. Early primates have recently been found in the late Cretaceous of North America. In the early Tertiary they are common in Europe as well.

Carnivora: cats, dogs, bears, weasels and the like. The first true carnivores appear in the Paleocene of North America.

Perissodactyla: horses, tapirs and other odd-toed ungulates. The earliest forms appear at the beginning of the Eocene in the Northern Hemisphere.

Artiodactyla:• cattle, deer, pigs and other even-toed ungulates. Like the odd-toed ungulates, they appear in the early Eocene of the Northern Hemisphere.

Rodentia: rats, mice, squirrels, beavers and the like. The first rodents appear in the Paleocene of North America.

Lagomorpha: hares and rabbits. This order makes its first appearance in the Eocene of the Northern Hemisphere.

Pholidota: pangolins. The earliest come from Europe in the middle Tertiary.

The fact that a given order of mammals is found in older fossil deposits in North America than in Europe or Asia does not necessarily mean that the order arose in the New World. It may simply reflect the fact that we know much more about the early mammals of North America than we do about those of Eurasia. All we can really say is that a total of 16 extant or extinct orders of mammals probably arose in the Northern Hemisphere.

Early Mammals of South America

The fragmentation of Gondwanaland seems to have started earlier than that of Laurasia. The rifting certainly had a much more radical effect. Looking at South America first, we note that at the beginning of the Tertiary this continent was tenuously connected to North America but that for the rest of the period it was completely isolated. The evidence for the tenuous early linkage is the presence in the early Tertiary beds of North America of mammalian fossils representing two predominantly South American orders: the Edentata (the order that includes today's ant bears, sloths and armadillos) and the Notoungulata (an order of extinct hoofed herbivores).

Four other orders of mammals are exclusively South American: the Paucituberculata (opossum rats and other small South American marsupials), the Pyrotheria (extinct elephant-like animals), the Litopterna (extinct hoofed herbivores, including some forms resembling

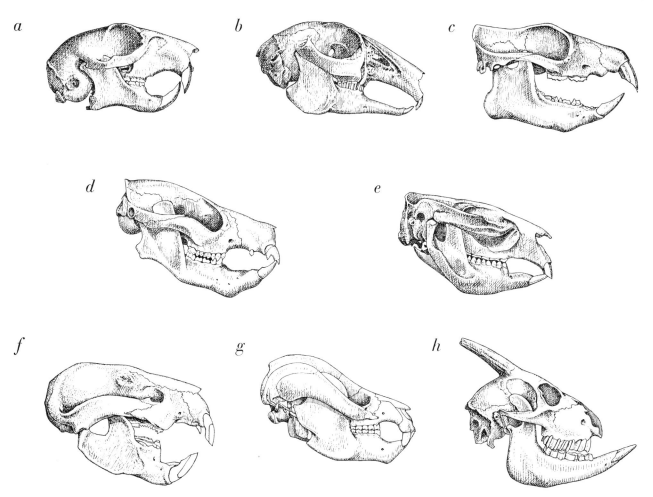

CHISEL-LIKE INCISORS, specialized for gnawing, appear in animals belonging to several extinct and extant orders in addition to the rodents, represented by a squirrel (*a*), and the lagomorphs, represented by a hare (*b*), which are today's main specialists in this ecological role. Representatives of other orders with chisel-like incisor teeth are an early tillodont, *Trogosus* (*c*), an early primate, *Plesiadapis* (*d*), a living marsupial, the wombat (*e*), one of the extinct multituberculate mammals, *Taeniolabis* (*f*), a mammal-like reptile of the Triassic, *Bienotherium* (*g*), and a Pleistocene cave goat, *Myotragus* (*h*), whose incisor teeth are in the lower jaw only.

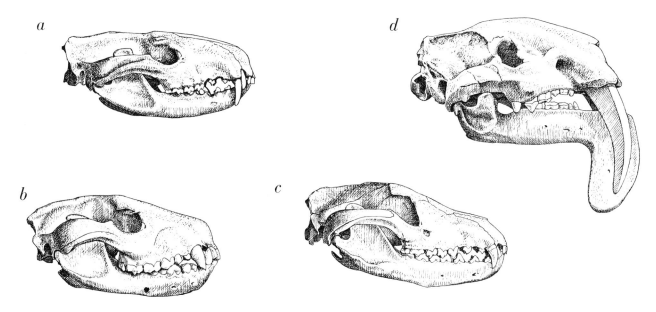

CARNIVOROUS MARSUPIALS, living and extinct, fill an ecological niche more commonly occupied by the placental carnivores today. Illustrated are the skulls of two living forms, the Australian "cat," *Dasyurus* (*a*), and the Tasmanian devil, *Sarcophilus* (*b*). The Tasmanian "wolf," *Thylacinus* (*c*), has not been seen for many years and may be extinct. A tiger-sized predator of South America, *Thylacosmilus* (*d*) became extinct in Pliocene times, long before the placental sabertooth of the Pleistocene, *Smilodon*, appeared.

horses and camels) and the Astrapotheria (extinct large hoofed herbivores of very peculiar appearance). Thus a total of six orders, extinct or extant, probably originated in South America. Still another order, perhaps of even more ancient origin, is the Marsupicarnivora. The order is so widely distributed, with species found in South America, North America, Europe and Australia, that its place of origin is quite uncertain. It includes, in addition to the extinct marsupial carnivores of South America, the opossums of the New World and the native "cats" and "wolves" of the Australian area.

The most important barrier isolating South America from North America in the Tertiary period was the Bolívar Trench. This arm of the sea cut across the extreme northwest corner of the continent. In the late Tertiary the bottom of the Bolívar Trench was lifted above sea level and became a mountainous land area. A similar arm of sea, to which I have already referred, extended across the continent in the region that is now the Amazon basin. This further enhanced the isolation of the southern part of South America.

Africa's role as a center of adaptive radiation is problematical because practically nothing is known of its native mammals before the end of the Eocene. We do know, however, that much of the continent was flooded by marginal seas, and that in the early Tertiary, Africa was cut up into two or three large islands. Still, there must have been a land route to Eurasia even in the Eocene; some of the African mammals of the following epoch (the Oligocene) are clearly immigrants from the north or northeast. Nonetheless, the majority of African mammals are of local origin. They include the following orders:

Proboscidea: the mastodons and elephants.

Hyracoidea: the conies and their extinct relatives.

Embrithopoda: an extinct order of very large mammals.

Tubulidentata: the aardvarks.

In addition the order Sirenia, consisting of the aquatic dugongs and manatees, is evidently related to the Proboscidea and hence presumably also originated in Africa. The same may be true of another order of aquatic mammals,

the extinct Desmostylia, which also seems to be related to the elephants. The one snag in this interpretation is that desmostylian fossils are found only in the North Pacific, which seems rather a long way from Africa. Nonetheless, once they were waterborne, early desmostylians might have crossed the Atlantic, which was then only a narrow sea, navigated the Bolívar Trench and, rather like Cortes (but stouter), found themselves in the Pacific.

Early Mammals of Africa

Thus there are certainly four, and possibly six, mammalian orders for which an African origin can be postulated. Here it should be noted that Africa had an impressive array of primates in the Oligocene. This suggests that the order Primates had a comparatively long history in Africa before that time. Even though the order as such does not have its roots in Africa, it is possible that the higher primates—the Old World monkeys, the apes and the ancestors of man —may have originated there. Most of the fossil primates found in the Oligocene

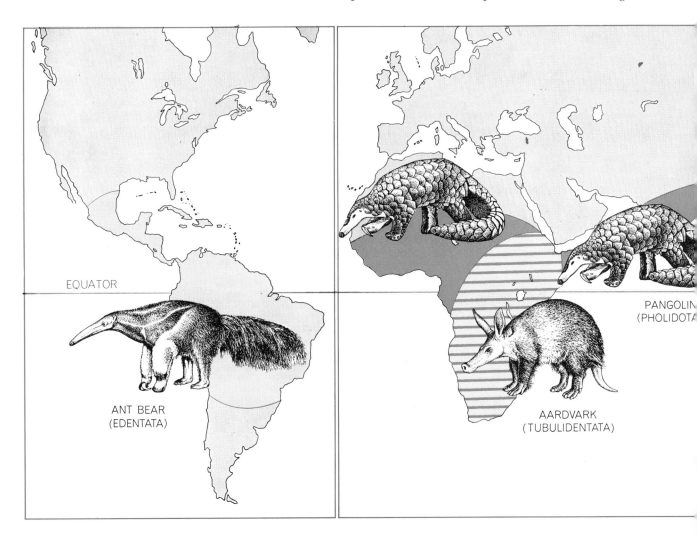

EQUATOR

ANT BEAR
(EDENTATA)

PANGOLIN
(PHOLIDOTA)

AARDVARK
(TUBULIDENTATA)

formations of Africa are primitive apes or monkeys, but there is at least one form (*Propliopithecus*) whose dentition looks like a miniature blueprint of a set of human teeth.

The Rest of Gondwanaland

We know little or nothing of the zoogeographic roles played by India and Antarctica in the early Tertiary. Mammalian fossils from the early Tertiary are also absent in Australia. It may be assumed, however, that the orders of mammals now limited to Australia probably originated there. These include two orders of marsupials: the Peramelina, comprised of several bandicoot genera, and the Diprotodonta, in which are found the kangaroos, wombats, phalangers and a number of extinct forms. In addition the order Monotremata, a very primitive group of mammals that includes the spiny anteater and the platypus, is likely to be of Australian origin. This gives us a total of three orders probably founded in Australia.

Summing up, we find that the three Laurasian continents produced a total

of 16 orders of mammals, an average of five or six orders per continent. As for Gondwanaland, South America produced six orders, Africa four to six and Australia three. The fact that Australia is a small continent probably accounts for the lower number of orders founded there. Otherwise the distribution—the average of five or six orders per subdivision—is remarkably uniform for both the Laurasian and Gondwanaland supercontinents. The mammalian record should be compared with the data on Cretaceous reptiles, which show that the two supercontinents produced a total of 12 or 13 orders (or adaptively distinct suborders). A regularity is suggested, as if a single nucleus of radiation would tend in a given time to produce and support a given amount of basic zoological variation.

As the Tertiary period continued new land connections were gradually formed, replacing those sundered when the old supercontinents broke up. Africa made its landfall with Eurasia in the Oligocene and Miocene epochs. Laurasian orders of mammals spread into Africa and crowded out some of the local forms, but at the same time some African mammals (notably the mastodons and elephants) went forth to conquer almost the entire world. In the Western Hemisphere the draining and uplifting of the Bolívar Trench was followed by intense intermigration and competition among the mammals of the two Americas. In the process much of the typical South American mammal population was exterminated, but a few forms pressed successfully into North America to become part of the continent's spectacular ice-age wildlife.

India, a fragment of Gondwanaland that finally became part of Asia, must have made a contribution to the land fauna of that continent but just what it was cannot be said at present. Of all the drifting Noah's arks of mammalian evolution only two—Antarctica and Australia—persist in isolation to this day. The unknown mammals of Antarctica have long been extinct, killed by the ice that engulfed their world. Australia is therefore the only island continent that still retains much of its pristine mammalian

fauna. [*see illustration on next page*].

If the fragmentation of the continents at the beginning of the age of mammals promoted variety, the amalgamation in the latter half of the age of mammals has promoted efficiency by means of a large-scale test of the survival of the fittest. There is a concomitant loss of variety; 13 orders of land mammals have become extinct in the course of the Cenozoic. Most of the extinct orders are island-continent productions, which suggests that a system of semi-isolated provinces, such as the daughter continents of Laurasia, tends to produce a more efficient brood than the completely isolated nuclei of the Southern Hemisphere. Not all the Gondwanaland orders were inferior, however; the edentates were moderately successful and the proboscidians spectacularly so.

As far as land mammals are concerned, the world's major zoogeographic provinces are at present four in number: the Holarctic-Indian, which consists of North America and Eurasia and also northern Africa; the Neotropical, made up of Central America and South America; the Ethiopian, consisting of Africa south of the Sahara, and the Australian. This represents a reduction from seven provinces with about 30 orders of mammals to four provinces with about 18 orders. The reduction in variety is proportional to the reduction in the number of provinces.

In conclusion it is interesting to note that we ourselves, as a subgroup within the order Primates, probably owe our origin to a radiation within one of Gondwanaland's island continents. I have noted that an Oligocene primate of Africa may have been close to the line of human evolution. By Miocene times there were definite hominids in Africa, identified by various authorities as members of the genus *Ramapithecus* or the genus *Kenyapithecus*. Apparently these early hominids spread into Asia and Europe toward the end of the Miocene. The cycle of continental fragmentation and amalgamation thus seems to have played an important part in the origin of man as well as of the other land mammals.

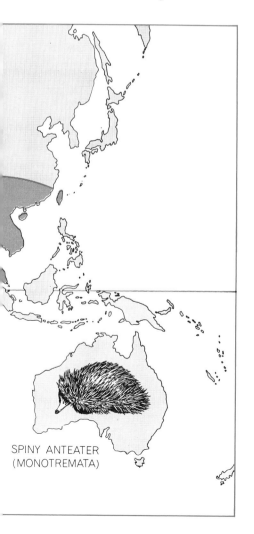

SPINY ANTEATER
(MONOTREMATA)

FOUR ANT-EATING MAMMALS have become adapted to the same kind of life although each is a member of a different mammalian order. Their similar appearance provides an example of an evolutionary process known as convergence. The ant bears of the New World Tropics are in the order Edentata. The aardvark of Africa is the only species in the order Tubulidentata. Pangolins, found both in Asia and in Africa, are members of the order Pholidota. The spiny anteater of Australia, a very primitive mammal, is in the order Monotremata.

122

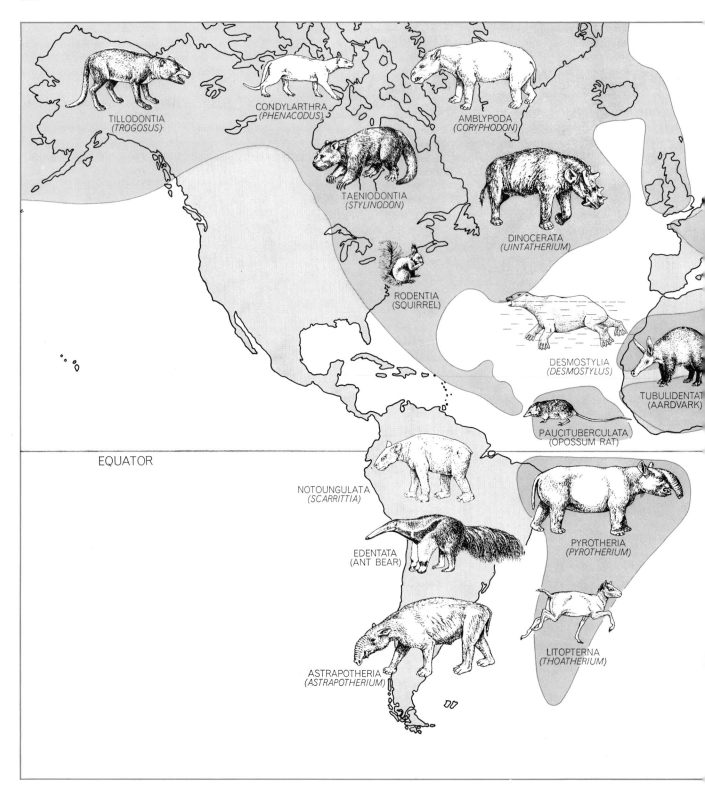

TILLODONTIA
(*TROGOSUS*)

CONDYLARTHRA
(*PHENACODUS*)

AMBLYPODA
(*CORYPHODON*)

TAENIODONTIA
(*STYLINODON*)

DINOCERATA
(*UINTATHERIUM*)

RODENTIA
(SQUIRREL)

DESMOSTYLIA
(*DESMOSTYLUS*)

TUBULIDENTAT
(AARDVARK)

PAUCITUBERCULATA
(OPOSSUM RAT)

EQUATOR

NOTOUNGULATA
(*SCARRITTIA*)

EDENTATA
(ANT BEAR)

PYROTHERIA
(*PYROTHERIUM*)

LITOPTERNA
(*THOATHERIUM*)

ASTRAPOTHERIA
(*ASTRAPOTHERIUM*)

CONTINENTAL DRIFT affected the evolution of the mammals by fragmenting the two supercontinents early in the Cenozoic era. In the north, Europe and Asia, although separated by a sea, remained connected with North America during part of the era. The

CHIROPTERA
(LITTLE BROWN BAT)

PHOLIDOTA
(PANGOLIN)

CREODONTA
(HYAENODON)

CARNIVORA
(WOLF)

PERISSODACTYLA
(BLACK RHINOCEROS)

PRIMATES
(RINGTAILED LEMUR)

ARTIODACTYLA
(GRANT'S GAZELLE)

LAGOMORPHA
(HARE)

MULTITUBERCULATA
(MESODMA)

INSECTIVORA
(WHITE-TOOTHED SHREW)

EMBRITHOPODA
(ARSINOITHERIUM)

SIRENIA
(MANATEE)

HYRACOIDEA
(ROCK CONEY)

MONOTREMATA
(SPINY ANTEATER)

DIPROTODONTA
(KANGAROO)

PROBOSCIDEA
(AFRICAN ELEPHANT)

PERAMELINA
(LONG-NOSED BANDICOOT)

free migration that resulted prevents certainty regarding the place of origin of many orders of mammals that evolved in the north. The far wider rifting of Gondwanaland allowed the evolution of unique groups of mammals in South America, Africa and Australia.

13

GEOSYNCLINES, MOUNTAINS, AND CONTINENT-BUILDING

ROBERT S. DIETZ
March 1972

A geosyncline is a long prism of sedimentary rock laid down on a subsiding region of the earth's crust. It has long been recognized that geosynclines are fundamental geologic units. Furthermore, it has been a dictum of geology that they eventually evolve into mountains consisting of folded sedimentary strata. The laying down of such sediments and their subsequent folding constitute a basic geologic cycle that requires a few hundred million years. Until recently the original nature of geosynclines has been inferred only by studying folded mountains. It was commonly believed that there are no nascent (unfolded) geosynclines in the world today, but this would defy another geologic dictum: that the present is the key to the past.

In recent years the study of marine geology has been revolutionized by the concept of plate tectonics, which holds that the earth's crust is divided into a mosaic of about eight rigid but shifting plates in which the continents are embedded and drift along as passive passengers. With this concept the evolution of ocean basins has been rather clearly resolved. The question arises: Must plate tectonics stay at sea, or is it also the prime mover of the geosyncline mountain-building cycle? In other words, can it account for the collapse of geosynclines and the growth of continents? I am among those who believe it can. Some notable advocates of this new concept of continental evolution are John Dewey and John M. Bird of the State University of New York at Albany, Andrew Mitchell and Harold Reading of the University of Oxford and William R. Dickinson of Stanford University.

When one examines the structure of ancient folded mountains, one finds that the classic geosyncline is divided into a couplet: two adjacent and parallel structures consisting of a eugeosyncline (true geosyncline) and a miogeosyncline (lesser geosyncline), often shortened to eugeocline and miogeocline. Now that the ocean floor is becoming better known, one need not look far to find an example of the geosynclinal couplet in process of formation. A probable example of a "living" eugeocline is the continental rise that lies seaward of the continental slope off the eastern U.S. Landward of the rise and capping the continental shelf is a wedge of sediments that becomes progressively thicker as it extends toward the shelf edge. This wedge seems to be a living miogeocline.

In dimensions and in the overall character of its rocks and stratigraphy the modern continental-rise prism closely matches typical ancient eugeoclines. It parallels the Atlantic seaboard for 2,000 kilometers, forming an apron 250 kilometers wide from the continental slope to the abyssal plain [*see illustrations on pages 128 and 129*]. Seismic studies reveal that the rise is the top of a huge plano-convex lens of sediments whose maximum depth is about 10 kilometers. The sediments are turbidites, deposited by the muddy suspensions known as tur-bidity currents. Such suspensions periodically cascade down submarine canyons and pour across the continental rise, depositing sedimentary fans that eventually coalesce into an apron. Turbidites consist of thin graded beds of poorly sorted particles of silt and sand in which coarse material is at the base and finer material is at the top. The gradation in particle size reflects the differential rate of settling from a single injection of muddy sand. Interlayered with the graded beds are fine clays (pelagites) that slowly settle from the overlying water as a "gentle rain" between major influxes of turbidity currents.

Collapsed eugeoclines in ancient folded mountains are similarly composed of thick and repetitive sequences of turbidites; these strata are usually termed flysch or graywacke. Mixed with the graywackes are thin limestones, ironstones and cherts formed from the skeletons of radiolarians, indicating that the sediments were deposited in deep water. True fossils are sparse, but many eugeoclinal sequences of the lower Paleozoic era contain graptolites: extinct plantlike animals that settled down from the surface.

Close examination of the graded beds also reveals what are called sole mark-

COLLISION OF CONTINENTS is depicted in this view of the Zagros Mountains in Iran along the Persian Gulf taken from the spacecraft *Gemini 12* in November, 1966. The mountains are uplifted folds of sedimentary strata, originally deposited as a geosyncline, whose cores have been exposed by erosion. The foldbelt has apparently been thrown up by the collision of the Arabian block, rotating counterclockwise, with the Eurasian block, rotating clockwise. Since the Arabian block is part of the African block, the folding represents the collision between Africa and Eurasia. The Zagros Mountains and the shallow Persian Gulf are both part of the Arabian block that extends to the Red Sea. The suture between the Arabian block and the Eurasian block is marked by a major thrust fault that passes through the upper right corner of the photograph just beyond the mountain chains.

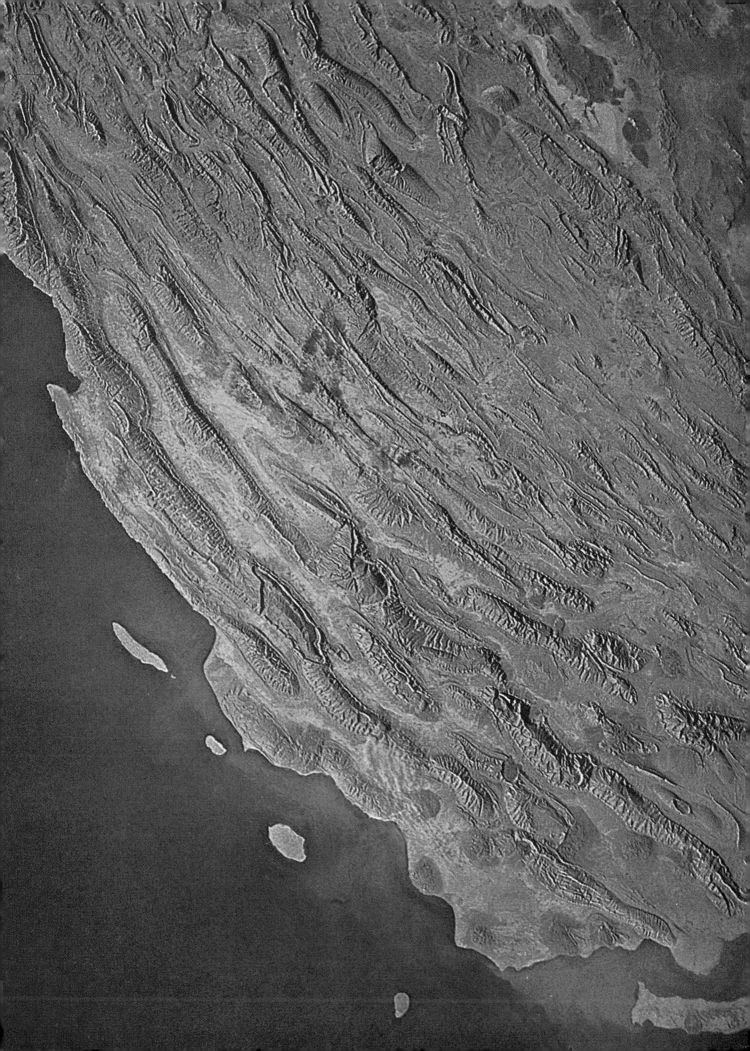

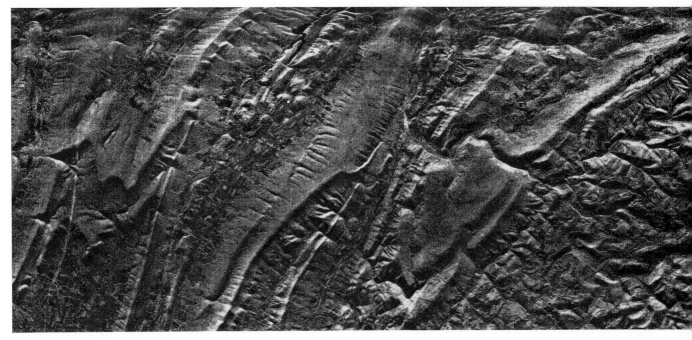

FOLDED APPALACHIAN MOUNTAINS in western Pennsylvania are depicted in this image produced by side-looking radar. The picture covers a region 25 miles long parallel to the Maryland border, centered approximately at 78 degrees 45 minutes west longitude. The picture is printed with north at the bottom so that the land-scape appears to be lighted from the top. (The illuminant, of course, is the radar beam transmitted from an airplane.) If the picture is inverted, features that are actually elevated appear depressed and vice versa. This image and the sequence of three views at the bottom of these two pages were made by the National Aeronautics

ings or "flysch figures," for example ripple marks of a kind that could have been produced by turbidity currents. There can be little doubt that most of these sequences are the uplifted and eroded remnants of former continental-rise prisms. The crystalline Appalachians, which are that part of the Appalachians lying seaward of the Blue Ridge Mountains and equivalent ranges to the north and south, bear the clear imprint of being a collapsed continental-rise prism

laid down in the early Paleozoic some 450 to 600 million years ago. The original prism has been much altered by intrusions and metamorphism.

The sedimentary wedge that underlies the coastal plain and continental shelf along the Atlantic seaboard appears to be an actively growing miogeocline. The wedge thickens as it progresses seaward, attaining a total thickness of between three and five kilometers along the shelf edge. Laid down

on a basement of Paleozoic rocks, the wedge is composed of well-sorted shallow-water sediments deposited during the past 150 million years under conditions much like those of today. The stratified beds exhibit characteristics indicating they were deposited across the continental shelf in alluvial plains, in lagoons, along shorelines and offshore. Taking into account expected changes in the pattern of sedimentation over geologic time, the present Atlantic marine

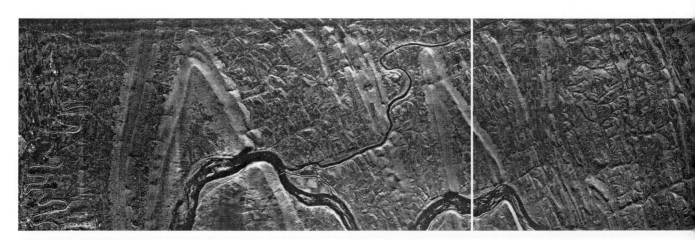

APPALACHIAN FOLDBELT north of Harrisburg, Pa., is an extension of the foldbelt shown at the top of the page. In these side-looking radar views north is at the right. The three pictures cover a distance of 75 miles from just south of Mechanicsburg to the vicinity of a town called Jersey Shore on the West Branch of Susque-hanna River. The Susquehanna River itself appears in the first frame at the left. The folded Appalachians were probably created in a late compressional stage of the collision between Africa and North America more than 450 million years ago, which caused "rugfolds" in the strata of sedimentary rock that formed part of a

and Space Administration in collaboration with the Remote Sensing Laboratory of the University of Kansas. The *K*-band radar system that produced the images was built by the Westinghouse Electric Corporation.

deposits closely resemble the ancient miogeoclinal foldbelts of the Paleozoic era and earlier. For example, the modern sedimentary wedge is much like the one found in the folded Appalachians of Pennsylvania. Both wedges are characterized by "thickening out," signifying that they grow steadily thicker toward the east before they abruptly terminate.

If the foregoing analysis is correct, one must conclude that geosynclines are actively forming along many continental

margins today: eugeoclines at the base of the continental slope and miogeoclines capping the continental shelves. It remains to be shown, however, that the crustal shifting associated with plate tectonics can convert these sedimentary prisms into the mountainous foldbelts that make up the fabric of the continents, mostly as ancient eroded mountain roots rather than as modern mountain belts. In order to examine this possibility we must first summarize some of the basic concepts of plate tectonics.

The approximately eight rigid but shifting plates into which the earth is currently divided are thought to be about 100 kilometers thick. Most of the plates support at least one massive continental plateau, often referred to as a craton. We can visualize the ideal plate as being rectangular, although only the plate supporting the Indian craton approaches this simple shape. Along one edge of a crustal plate there is a subduction zone, usually marked by a trench, where the plate dives steeply into the earth's mantle, attaining a depth as great as 700 kilometers before being fully absorbed into the mantle. On the opposite side of the plate from the subduction zone is a mid-ocean rift, or pull-apart zone. As the rift opens, the gap is quickly healed from below by the inflow of liquid basalt and quasi-solid mantle rock. The other two opposed sides of the plate, connecting the rifts to the trenches, are shears called transform faults.

Hence three types of plate boundary are possible: divergent junctures (the mid-ocean rifts where new ocean crust is created), shear junctures (the transform

faults where the plates slip laterally past one another, so that crust is conserved) and convergent junctures (trenches where two plates collide, with one being subducted and consumed). Only the last of the three, the convergent juncture, can help to explain how the sedimentary prism of a submarine geosyncline might be collapsed into a folded range of mountains. As the plate carrying a prism collides with a plate carrying a continental craton one would expect the prism to be compressed into folds. Thrusting and crustal thickening would follow, assisted by isostatic forces that act to keep adjacent crustal masses in balance. Such forces would cause the collapsed prism to be uplifted. The entire process would be accompanied by the generation and intrusion of magma, together with extensive metamorphism of the crustal rocks.

A grand theme of plate tectonics is that ocean basins are not fixed in size or shape; they are either opening or closing. Today the Atlantic Ocean is opening and the Pacific Ocean is closing. The drifting of the continents is another theme; every continent must have a leading edge and a trailing edge. For the past 200 million years the Pacific coast of North America has been the leading edge and the Atlantic coast the trailing edge. The trailing margin is tectonically stable, and since the continental divide is near the mountainous Pacific rim, most of the sediments are ultimately dumped into the Atlantic Ocean, including the Gulf of Mexico. Therefore it is primarily along a trailing edge that the great geosynclinal prisms are deposited.

Consider, however, what would hap-

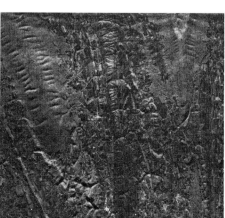

Paleozoic geosyncline *(see illustration on page 130).* After folding the region was eroded to a level plain and then uplifted. The modern mountains were subsequently etched out according to the hardness of the various strata. Thus the ridges are composed mainly of dense sandstone and can be either synclines (troughs) or anticlines (arches). The V-shaped chevrons in the first frame are synclines that plunge to the northeast. The Susquehanna River established its course when the entire region was reduced to a level surface, so that its course has been superimposed on the folded structure, thereby cutting directly across the folds and creating water gaps.

pen if, with the changing patterns of plate motions, a subduction zone (a trench) were created along a former trailing edge, forming a new plate boundary. The Atlantic would be transformed into a closing ocean with its geosynclinal prisms riding toward the trench. The continental margin and the trench would eventually collide, collapsing the eugeocline into a contorted mountainous foldbelt and also folding the miogeocline to a much lesser extent. Before that happened the continental margin would encounter and incorporate an island arc, similar to the island arcs found along the perimeter of the western Pacific. These arcs are created by tectonic and magmatic activity triggered by the plunging crustal plate. It is also quite possible that the Atlantic Ocean would close entirely, causing North Africa to collide with eastern North America. The collision of India with the underbelly of Asia, throwing up the Himalaya rampart, would be a present-day analogy. One can imagine many possible scenarios, depending on the geometry of plate boundaries and other variables.

The creation of a eugeoclinal foldbelt is of course considerably more than simply the accordion-like collapse of a continental-rise prism. The foldbelt is sheared into thrust faults and the landward edge of the eugeocline is common-

ly thrust onto the adjacent miogeocline. The descending crustal plate is not entirely consumed within the earth's hot mantle, with the result that low-density magmas buoyantly rise and invade the eugeocline. This leads to intrusions of granodiorite (a granite-quartz rock) and the growth of volcanic mountains consisting of andesite (the rock characteristic of the Andes). This lava is highly explosive because it is charged with water sweated out of the descending plate. Magma is not generated from the plunging lithosphere until it has reached a considerable depth. As a result the eugeocline can be subdivided into two parallel geologic belts. Toward the sea one finds sedimentary rock transformed at high pressure and low temperature; farther inland the sedimentary rock has been altered predominantly at low pressure and high temperature by the numerous intrusions of magma. From the new marginal mountain range, delta and river deposits sweep back across the continent, covering the miogeocline with a suite of continental shales and conglomerates collectively called molasse.

The concept that the geosynclinal cycle is controlled by plate tectonics provides some new answers to old questions about geosynclines. For example, is mountain-building periodic and world-

wide or is it random in space and time? The answer must be both yes and no. On the one hand, the crustal plates are highly intermeshed; the drift of any one plate has global repercussions, giving rise to synchronous mountain-building. Any brief interval of rapid plate motion would also be one of widespread mountain-building. On the other hand, the rate of plate convergence is highly dependent on the latitudinal distance from the relative pole of rotation of that plate and on the particular geometry of the plate boundary.

A law of plate tectonics states that sea-floor spreading (injection of new ocean crust) proceeds at right angles to a rift; the crustal plate, however, may be subducted into a trench at any angle. The rate of subduction and the attendant distorting of the crust therefore vary greatly from place to place, as can be observed on the perimeter of the Pacific today. Thus it would seem that although mountain-building over the span of geologic time may reach crescendos, it must also be continuous and random.

The plate-tectonic version of the geosynclinal cycle predicts that miogeoclines are ensialic, or laid down on continental crust (sial), whereas eugeoclines are ensimatic, or deposited on oceanic crust (sima). This differs from the earlier view that all geosynclines are ensialic,

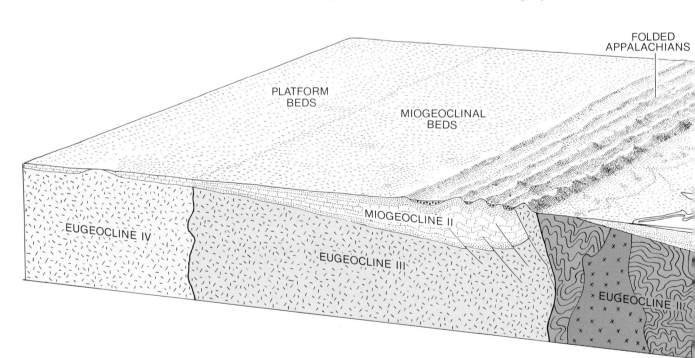

SUCCESSION OF EUGEOCLINES underlies nearly all North America below the relatively undisturbed cover beds. These contorted and intruded prisms constitute the fundamental fabric of continents, known as the basement complex. A new geosynclinal couplet is being deposited today. It consists of a miogeocline (lesser geosyncline) of shallow water beds that caps the coastal plain and continental shelf paralleled by a eugeocline (true geosyncline) that is formed at the base of the continental slope by detritus

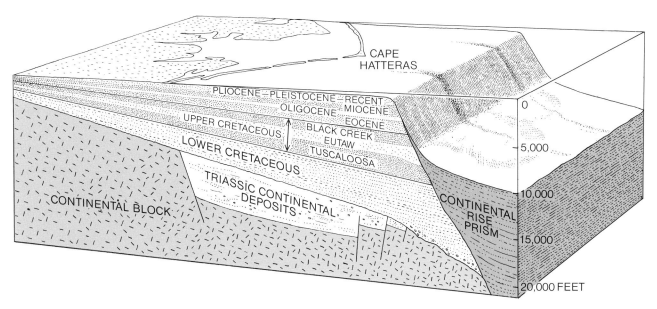

LIVING GEOSYNCLINAL COUPLET off the Atlantic coast of the U.S. consists of a miogeocline, strata laid down on the shallow continental shelf during the past 150 million years, and a eugeocline prism *(dark color)*, consisting of thin beds of sand and mud de-posited by turbidity currents flowing down the continental slope. The material in the Triassic basin represents continental deposits laid down before the foundering of the continental margin under tension 190 million years ago, prior to opening of Atlantic Ocean.

which is certainly incorrect. Early investigators observed that a granitic basement is invariably present under miogeoclines and evidently reasoned that a similar sialic basement, although it was unseen, must also be present under eugeoclines. A collapsed eugeocline is as thick as the continental plate, about 35 kilometers, so that its basement is beyond the depth of even the deepest boreholes. We can infer that the ultimate basement is simatic, however, by observing detached fragments that are caught up in the contorted mélange of the eugeoclinal pile. These fragments include samples of oceanic crust (for example radiolarian cherts and sodium-altered lavas) and upper-mantle rocks (for example serpentinites and peridotites).

The ensimatic location of eugeoclines can also account for their tectonic style. They are tightly folded, faulted, tumbled and dynamically metamorphosed into an almost unmappable mélange. This contorted state is understandable, since the ocean floor is shearing under the eugeocline and thrusting the sedimentary pile against the continental slope. Extensive tectonic thickening and interleaving must occur before the pile will rise to mountainous altitudes. On the other hand, the miogeocline beds are protected by the stable continental slab, so that they are simply thrown into a series of loose, open, ruglike folds.

It now seems amusing to recall that 19th-century geologists, using a wrinkled apple as an analogy, interpreted folded mountains to mean that the earth was shrinking. Today it seems clear that eugeoclines are deposited at the edge of a continent on oceanic crust seaward of the continental slope, so that folded mountains really show that the continents are growing larger through marginal accretion. Mountain-building is therefore evidence of an even more fundamental geologic process: the growth of continents. The continents grow not as a layer cake but as a craton that is divided vertically into zones with an old nucleus and young margins.

An important aspect of geosynclines requiring explanation is that they are laid down on foundations that are continuously subsiding. This aspect is par-

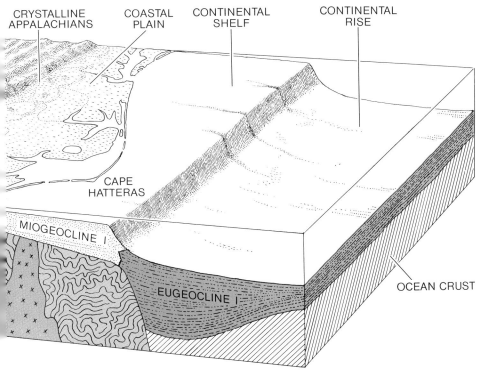

washed over the shelf edge. If at some future time the sea floor were to thrust against the continent, the modern eugeocline *(I)* would collapse into a new foldbelt like the earlier ones. The hypothetical mechanism that creates foldbelts is shown on page 131. This diagram and others are based on drawings by the author's colleague John C. Holden.

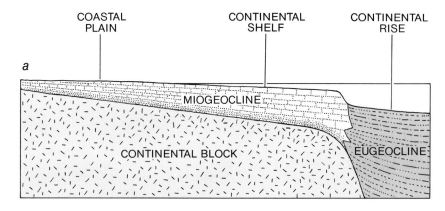

a

COASTAL PLAIN — CONTINENTAL SHELF — CONTINENTAL RISE

MIOGEOCLINE

CONTINENTAL BLOCK

EUGEOCLINE

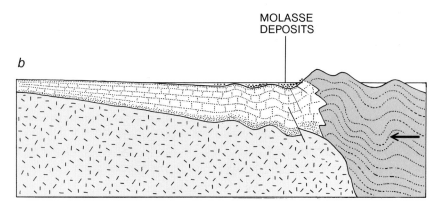

b

MOLASSE DEPOSITS

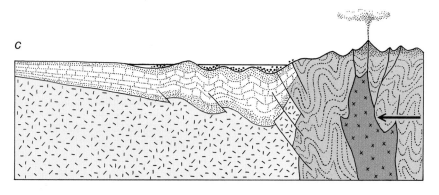

c

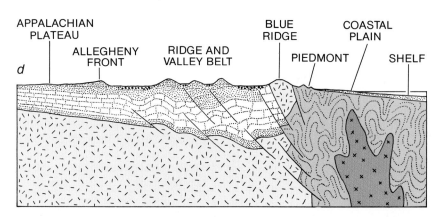

d

APPALACHIAN PLATEAU

ALLEGHENY FRONT

RIDGE AND VALLEY BELT

BLUE RIDGE

PIEDMONT

COASTAL PLAIN

SHELF

CRUMPLING OF EARLIER GEOSYNCLINAL COUPLET, apparently laid down in late Precambrian and early Paleozoic time more than 450 million years ago, produced the Appalachian foldbelt. The four-part sequence shows how the miogeocline, or western part of the geosynclinal couplet, was folded into the series of ridges between the Blue Ridge line and the Allegheny front. The eugeocline, altered by heat, pressure and volcanism, formed a lofty range of mountains, now almost completely eroded, east of the Blue Ridge line.

ticularly evident in miogeoclines, which can attain a total thickness at their seaward edge of five kilometers even though they are entirely composed of beds deposited in shallow water. This phenomenon is nicely accounted for by plate tectonics: the margins of rift oceans inherently have, as one geologist has expressed it, a "certain sinking feeling."

Let us take as an example the Atlantic Ocean between the U.S. and the bulge of Africa. This new ocean basin was created about 180 million years ago by the insertion of a spreading rift that split North America away from Africa [see illustration on opposite page]. Attendant swelling of the mantle arched the continents upward along the rift line by about two kilometers. Erosion then beveled the raised edges, thinning the margins of the two continental plates.

A modern example of crustal arching associated with incipient rifting of the crust can be observed in the high dorsal of Africa from Ethiopia southward. The Red Sea provides a more advanced stage of a newly opening ocean basin. Along the flanks of this linear trough crustal arching has stripped away young rocks, exposing "windows" of Precambrian basement.

With the insertion of new oceanic crust by sea-floor spreading, the ocean grew ever wider. In the process the continental edges subsided, as is demonstrated by the sloping flanks of the mid-Atlantic ridge today. When the ocean was smaller, the continental edges had to ride down a similar slope. Eventually the inflated mantle under the ridge reverts to normal mantle, but this takes 100 million years or more. Therefore as a geosyncline is laid down on the trailing edges of a drifting continent it slowly subsides for reasons external to the sedimentary deposit itself.

Additional subsidence, however, is caused by the steadily growing mass of the sedimentary apron, which must be isostatically compensated because the earth's crust is not sufficiently strong to sustain the load. For every three meters of sediment deposited the crust sinks about two meters. This crustal failure, however, is spread over a large geographic area, so that the growth and subsidence of a huge continental-rise prism causes a sympathetic downward flexing of the adjacent continental margin. As the continental shelf slowly tilts, wedges of shallow-water sediments are deposited.

Over the millenniums, as the shoreline transgresses and regresses repeatedly across the shelf, a large composite

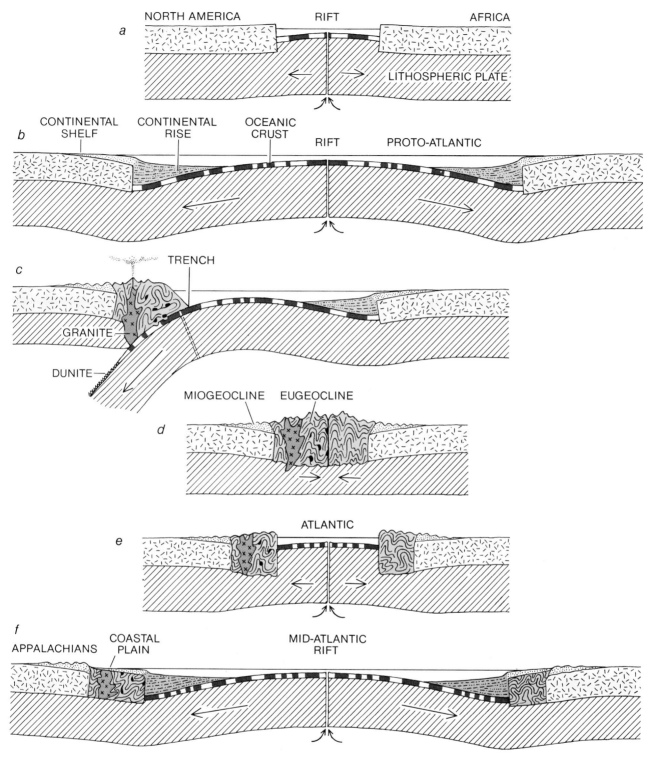

MECHANISM OF CRUMPLING that produced the Appalachian foldbelt is depicted on the hypothesis that the Atlantic Ocean has opened, closed and reopened. In the late Precambrian (*a*), North America and Africa are split apart by a spreading rift, which inserts a new ocean basin. By the process of sea-floor spreading (*b*) the ancestral Atlantic Ocean opens. New oceanic crust is created as the plates on each side move apart. As the crust cools, its direction of magnetization takes the sign of the earth's magnetic field; the field periodically reverses, and the reversals are represented by the striped pattern. On the margin of each continent sediments produce the geosynclinal couplet: miogeocline on the continental shelf, eugeocline on the ocean floor itself. The ancestral Atlantic now begins to close (*c*). The lithosphere breaks, forming a new plate boundary, and a trench is produced as the lithosphere de-

scends into the earth's mantle and is resorbed. The consequent underthrusting collapses the eugeocline, creating the ancient Appalachians. The eugeocline is intruded with ascending magmas that create plutons of granite and volcanic mountains of andesite. The proto-Atlantic is now fully closed (*d*). The opposing continental masses, each carrying a geosyncline couplet, are sutured together, leaving only a transform fault (*vertical black line*). The shear contains squeezed-up pods of ultramafic mantle rock. Sediments eroded from the mountain foldbelt create deltas and fluvial deposits collectively called molasse. North America and Africa were apparently joined in this way between 350 million and 225 million years ago. About 180 million years ago (*e*) the present Atlantic reopened near the old suture line. Today (*f*) the central North Atlantic is opening at the rate of three centimeters per year, creating new geosynclines.

megawedge of shallow-water sediments caps the shelf. The abundant supply of sediment usually ensures that the top of the prism is maintained close to sea level. Excess detritus bypasses the shelf, is temporarily dumped on the continental slope and is then carried onto the continental rise by turbidity currents. The shelf edge and the continental-rise prism comprise a couplet within which there is constant interplay.

Like the sedimentary wedge under the Atlantic coastal plain today, the early Paleozoic Appalachian miogeocline thickens in the seaward direction. The abrupt termination of this miogeocline was long a mystery to early geologists who mapped it. They suggested that a missing seaward limb had been thrust upward and completely eroded away or that it had foundered into an ancient oceanic basin. This hypothetical land mass of Appalachia was the geological equivalent of the legendary Atlantis. The wedgelike structure of the existing continental-plain prism provides a satisfactory solution to the puzzle: the hypothesized seaward limb never existed. We now see that the thickening out of sedimentary deposits at the shelf edge is a normal mode of sedimentation. One way in which this may happen is that reefs of coral and algae build up along the margin of the continental shelf, creating a carbonate dam behind which other shelf sediments accumulate.

The mechanism of building continents by the peripheral accretion of collapsed continental rises seems also to ensure that the sedimentary deposits become dry land. (We take for granted that continents are above sea level, but it should be remembered that the mid-ocean ridge system, which approaches the continents in importance as a topographic feature, almost never rises above the sea surface.) The sedimentary apron gradually thickens until it approaches the height of the continental slope (about five kilometers), but upward growth ceases once the slope is completed and sea level is attained.

As we have seen, however, isostasy is at work, causing the oceanic crust to subside under the sedimentary load. The result is that a fully developed sedimentary prism attains an overall thickness of about 15 kilometers. When the prism is subsequently collapsed into a eugeosynclinal foldbelt, it becomes thicker still. The attendant metamorphism and granitic intrusion (which increases the total mass of rock) give rise to a monolithic structure that is more than 35 kilometers thick, thicker than a continental plate.

Thus new foldbelts not only rise above sea level but also throw up rugged mountain ranges.

The hypothesis that geosynclines are deposited along a continental margin and then crushed against the continent as a result of plate tectonics seems to explain satisfactorily how geosynclines are transformed into folded mountains. The close relation between eugeoclines and foldbelts is not one of cause and effect but a simple consequence of location: geosynclines are laid down along continental margins and such margins are the locus of interaction between continents and subduction zones.

In spite of the vast span of geologic time and the rigors of erosion, the continents remain in a good state of health. We can predict that they always will be: detritus lost to the oceans is eventually carried back to the continents and collapsed into accretionary belts that also incorporate new igneous rock. Although the earthquakes that punctuate mountain-building are sometimes disastrous to man's culture, they are acts of continental construction. The great flood—the complete inundation of the erosionally leveled continents—will always threaten but will never come to pass.

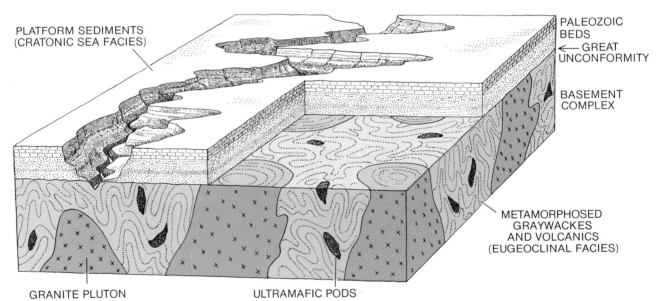

PLATFORM SEDIMENTS
(CRATONIC SEA FACIES)

PALEOZOIC BEDS
← GREAT UNCONFORMITY

BASEMENT COMPLEX

METAMORPHOSED GRAYWACKES AND VOLCANICS (EUGEOCLINAL FACIES)

GRANITE PLUTON   ULTRAMAFIC PODS

DEEP FABRIC OF CONTINENTS, the basement complex, is the fundamental rock unit of the continental plateaus, or cratons. This complex is usually obscured from view by miogeoclinal beds or by the coating of shallow-sea deposits that have invaded the continents from time to time. (Hudson Bay is a modern example of such an invading shallow sea.) Long a puzzle to geologists, the basement is composed of eugeoclinal foldbelts that have undergone intensive folding, metamorphism and intrusion. Geologists once thought that these "roots of mountains" indicated that the earth had contracted while cooling. The folds were likened to the skin of a dried apple. The present interpretation is that the eugeoclinal facies was laid down on the ocean floor and subsequently was crumpled against the continental margin, building up an onion-like vertically zoned craton. On this view the continents have grown larger rather than smaller with time. Moreover, the basement complex need not be Archean (composed of the oldest rocks) as formerly supposed, because the high degree of metamorphism does not necessarily indicate great antiquity and repeated mountain-building events. Instead it reflects the intensity of the collapsing process; once accreted, the foldbelt is not usually mobilized again. It has long been known that the granites of the basement complex are always younger than the metamorphosed sediments they intrude. On the other hand, the included pods of ultramafic rocks are always older, since they are detached fragments of the oceanic foundation on which the eugeocline was deposited. If geologists ever find "the original crust of the earth," it will be one of these pods within the oldest foldbelts.

14

THE AFAR TRIANGLE

HAROUN TAZIEFF

February 1970

In the northeastern part of Ethiopia, at the juncture of the Red Sea and the Gulf of Aden, lies a region known as the Afar triangle. It is a wild and rugged country, featured by below-sea-level deserts, towering escarpments, fissures, volcanoes and craters. Few men have explored the region, and until recently it was terra incognita as far as its geology and even its exact geography were concerned. Now, however, the discovery of detailed evidence for the drift of continents and the growth of oceans has focused considerable interest on the Afar triangle [see "The Origin of the Oceans," by Sir Edward Bullard; SCIENTIFIC AMERICAN, September, 1969]. The triangle seems to be a focal point for new oceans in the making. What is more, whereas elsewhere the process that is producing continental separation is hidden in the depths of the ocean, here we can see it taking place in direct view on dry land.

The theory that the earth's present continents were once united in a great land mass and have gradually drifted apart is now generally accepted. Evidence collected over the past 10 years, mainly by exploration of the worldwide system of ridges running along the middle of the oceans, has outlined a convincing picture of how the continents were separated [see the article "The Confirmation of Continental Drift," by Patrick M. Hurley, beginning on page 57]. In brief, the process seems to be as follows. The material of the continents is a layer of comparatively light sialic rock, resting on a denser basaltic magma underneath. Stresses in the earth's crust may crack the sialic layer, producing faults and fissures that can be as much as 20 meters wide. Then molten magma wells up into the fissure, sometimes spilling out over the surface. The magma hardens into solid rock, thus holding apart the separated sialic blocks. Over long periods of time the same stresses create new fissures parallel to the old ones; these fissures too are filled with magma. Examination of the oceanic ridge has shown that it is composed of parallel strips of hardened magma, indicating that the crust was repeatedly fissured along the axis of the ridge and that the continental blocks thus moved farther and farther apart. The upwelled strips of basalt are distinguishable from one another by differences in the direction of their magnetic polarity, which

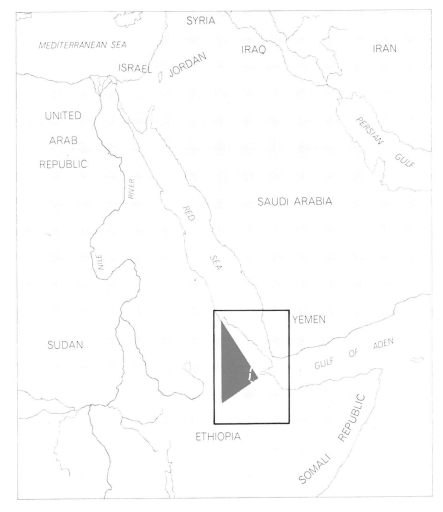

AFAR TRIANGLE *(color)* is in northeastern Ethiopia where the Red Sea rift, the Carlsberg Ridge of the Indian Ocean and the Rift Valley system of East Africa meet. Its northern section was once submerged, and part is still below sea level *(see map on pages 134–135).*

reflect changes in the earth's magnetic field with time.

The overall conclusion from studies of the mid-oceanic ridges is that the continents have been moving apart at an average rate of a few centimeters per year. In the South Atlantic, for instance, it appears that the ocean has been widening at the rate of 2.5 centimeters per year, which indicates that the South American continent began to separate from Africa approximately 180 million to 200 million years ago.

Recent worldwide explorations of the ocean bottoms have shown that a rise running along the middle of the Indian Ocean, known as the Carlsberg Ridge, has a branch extending into the Gulf of Aden. The Red Sea bottom similarly has an axial ridge with physical properties like those of the oceanic ridges. The Gulf of Aden and Red Sea rifts, which are perpendicular to each other, meet in the Afar triangle. And the same region also lies at the northern end of the system of rift valleys that runs down the eastern side of the African continent. Once these facts were recognized, it was suddenly realized that the largely uncharted Afar triangle might offer an extraordinary opportunity for investigating the origin of new oceans.

The Afar triangle is one of the world's most forbidding regions. In addition to the fact that its terrain is all but impassable, the area is extremely hot; we were to find that the temperature rises to as high as 134 degrees Fahrenheit in the shade in summer and 123 degrees in winter. The region is inhabited only by nomadic tribes of fierce repute; the young warriors are said to mutilate male victims to offer trophies to their women, and they have been known to massacre armed parties for their weapons. Several exploring parties in the 19th century were slaughtered by the tribesmen.

Of the expeditions that had explored the Afar triangle before we began to survey it in 1967, the best-known was one carried out in the spring of 1928 by two Italians, Tullio Pastori and G. Rosina, a British mining engineer, L. M. Nesbitt, and half a dozen Ethiopians. The expedition's leader was Pastori, a hardy ore prospector who earlier had explored various parts of the region (and who, at the age of 60, in 1943 escaped from a British prisoner-of-war camp in Kenya and made his way on foot all the way to Alexandria on the Mediterranean coast). Nesbitt, who wrote a report on the 1928 expedition, vividly described the difficulties and dangers the expedi-

tion had encountered. It was apparently the first journey along the entire length of the Afar triangle, and Nesbitt's information provided a basis for the only detailed maps of the region that were available before our expeditions. The 1928 party was not equipped, how-

ever, to undertake a geological survey.

In 1967 we organized a team of specialists to study the geology of the Afar triangle with all possible thoroughness. Our group includes investigators from several universities in Italy, France and the U.S. It consists of three petrologists

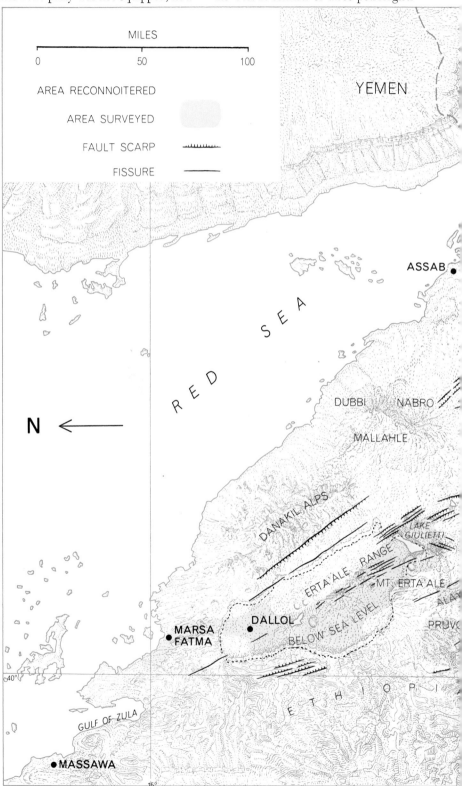

GEOLOGICAL SURVEY of the Afar triangle was conducted by the author and his associates beginning in 1967. In this map north is at the left and east at the top. The most inten-

(Giorgio Marinelli, Franco Barberi and Jacques Varet), four geochemists (Giorgio Ferrara, Sergio Borsi, J. L. Cheminée and Marino Martini), one tectonic geologist (Gaetano Giglia), two students of the geology of recent times (Hugues Faure and Colette Roubet), a geophysicist (Guy Bonnet), an oceanographer (Enrico Bonatti of the University of Miami) and a volcanologist (myself).

We have now completed three expeditions to the Afar triangle, all during the comparatively cool winter season (in 1967–1968, 1968–1969 and 1969–1970). Our party was unarmed and was not molested by the local people. For our explorations we have had the invaluable help of a helicopter, without which it would simply have been impossible to do serious, comprehensive fieldwork in that rugged country. So far we have

sive work was done in and near the area below sea level, between the Danakil Alps and the Ethiopian escarpment. In addition the group scouted widely by helicopter, landing more than 1,000 times to gather rock samples for later analysis and to study formations.

MASSIVE FAULT BLOCKS characterize the terrain of the Afar triangle. The aerial photograph above shows a region near 12 degrees north latitude, where faulting has caused huge segments of surface to subside below the level of the surrounding plateau. The depressed structures are whitened in places by salt or gypsum.

STAIRSTEP LANDSCAPE of the eastern shore of Lake Giulietti (*below*) is further evidence of the extensive downfaulting that has shaped much of the Afar triangle. Such long, depressed blocks, bordered by fault zones, are called graben structures by geologists; the graben structures here are among the world's most spectacular.

spent a total of 13 weeks in the field and have carefully mapped the geological structures and petrology of some 12,000 square miles in the northern half of the triangle. We have crisscrossed this entire area with the helicopter and have landed at more than 1,000 places to sample rocks and examine the tectonic structures on the ground.

On its northeast side the Afar triangle is separated from the Red Sea by a series of heights produced by deformations of the earth crust at faults and by volcanic action. Reading from north to south, these include a tectonic horst (upraised block) called the Danakil Alps, a mountain range formed by sedimentary and intrusive rocks, a volcanic massif composed of an active volcano and three volcanic piles crowned with calderas (craters of collapsed volcanoes) and, at the southern end, another mountainous horst. On the south side the triangle is bounded by the tall Somali scarp, from 4,900 to 6,500 feet high. On the west, the third side of the triangle, stands the huge Ethiopian escarpment, towering in some places to more than 13,000 feet. Here clifftops stand higher above the valley floor below them than anywhere else in the world. As for the floor of the triangle itself, it rises gently from about 400 feet below sea level near its northern end to more than 3,300 feet above sea level at the Somali end some 300 miles south.

What is one to make of this strange and spectacular landscape? Until very recently most geologists believed the Afar triangle was a funnel-shaped widening, produced in some unexplained manner, of the Great Rift Valley of East Africa. Our studies of the region's geology have led us to a completely different interpretation. The facts, as we have observed them on the scene, indicate that the floor of the triangle is actually a part of the Red Sea! In fact, the triangle from its northern apex down to the Ghubbet al Kharab at the western end of the Gulf of Aden and the Gulf of Tadjoura is a southwestern continuation of the central trough of the Red Sea that fades out close to 15 degrees south latitude. The tectonic trends of the Red Sea are evident and are geologically active throughout the area. There are none of the trans-rift structures or other formations that were once believed to account for the Afar triangle. It appears that as recently as some tens of thousands of years ago at least the northern half of this area was covered with seawater, with only the Danakil Alps and the high volcanoes standing above water as islands, and that most of the land has since been raised above sea level by tectonic uplifts through earthquakes, volcanic action and the rise of basaltic magma that filled the fissures.

The evidence that the Afar triangle is a part of the Red Sea floor, not some bizarre widening of the Ethiopian rift, can be seen on every hand. To begin with, we found that the observable facts contradicted other explanations of the region's topography. It had been suggested, for example, that the lofty escarpment on the west side of the triangle was produced by a downfolding of the high plateau, followed by erosion of the resulting hillside. Our observations turned up three important objections to this idea: (1) the blocks of faulted crust in the lower part of the escarpment are tilted westward—in the direction opposite to what would be expected in a downfold; (2) the supposed erosion should have deposited a vast amount of sediment in the triangle's closed basin, but almost none was found there; (3) the basin is filled with more than 3,000 feet of evaporites (salt formations deposited by the evaporation of seawater), and the basin's wall plunged down below this material at a steep angle, again indicating that the western boundary of the triangle was formed by slippage of the crust along a fault rather than by downwarping.

It had also been suggested that south of the Danakil Alps a belt of big calderas, apparently running north and south, was a continuation of the main Ethiopian rift and was a major active feature of the entire region. Field investigations show that no north-south belt exists, and that all the big calderas are located on a graben (a depressed section of crust) running north-northeast and south-southwest. This observation is an important one, because it again demonstrates that the Afar triangle is a part of the Red Sea and not an extension of the Ethiopian rift.

We found innumerable signs that the topography of the Afar triangle has been created by violent events that have occurred in very recent times, geologically speaking, and are still in progress. The entire northern half of the triangle shows clear evidence of extensive faulting of the crust and active crustal movement along the faults. North of latitude 13 degrees 10 minutes north all these faults are aligned along the axis of the triangle in the north-northwest to south-southeast direction [see illustration on pages 134 and 135]. South of 13 degrees 10 minutes north down to the Ghubbet al Kharab (11 degrees 30 minutes north) the same direction prevails, although there are also many faults and fissures running northwest-southeast and east-west. Along much of this part of the triangle the evidences of crustal movement are in plain view as wide-open fissures, horsts and grabens that form a classic graben structure of steps down the sides of a major depression. Even where the graben structure is now hidden under deposits of volcanic material and evaporites (but is still detectable by geomagnetic mapping), the fault axis is shown visibly by potash salt domes, explosion craters, boiling springs and other signs of volcanic activity below the surface. These eruptions follow the same line in the north-northwest to south-southeast direction.

This direction is precisely that of the Red Sea, so that the whole of the northern part of the Afar triangle can be regarded as part of the sea. All the evidence suggests that the waters of the Red Sea extended south-southeastward as far as the Somali scarp in the geologically recent past and that the present absence of water in the Afar triangle is only a temporary phase in the development of the ocean.

That the fissuring and displacements of the crust are going on actively at the present time is shown by several signs we were able to observe. Here and there we found fresh faults cutting through very young structures, such as alluvial fans and cones of active volcanoes. The volcano called Erta'Ale, the most active in all East Africa, has its main cone sliced by tectonic fissures that are parallel to the Red Sea axis. This is an exceptional phenomenon; volcanic fissures are usually not parallel but radial. We ourselves actually witnessed a significant event during our study: an earthquake in the middle of the depression on March 26, 1969, produced an appreciable slippage of the crust down a fault in the north-northwest to south-southeast direction.

Further evidence that the Afar triangle is actually a scene of oceans in the making came from the examination of the rocks themselves. Our samples gave every indication that the triangle's central trough contains no sialic (that is, continental) rock. The rocks, all very young, of the Erta'Ale volcanic range, which runs parallel to the Red Sea central trough, are preponderantly basaltic and typical of the rocks of oceanic ridges. From analysis of more than 100

FLAT-TOPPED VOLCANO, Mount Asmara is composed of shards of volcanic glass such as are formed during underwater volcanic explosions. It resembles the numerous guyots, or submerged ocean- ic mountains, whose level summits are usually attributed to wave erosion. Because Mount Asmara was formed under water, it may be that a flat top is instead a feature common to all such volcanoes.

CINDER CONE near Lake Giulietti was built up at a point that overlies one of the innumerable fault lines found in the Afar tri- angle. A subsequent horizontal shift of one fault block has moved the far half of the cinder cone 100 meters ahead of the near half.

specimens from this range we estimate that its composition is more than 90 percent basalts, the rest consisting of varieties of volcanic rock: dark trachytes (8 percent) and rhyolites (.5 percent). Chemically these rocks all show an evolutionary relationship, forming a practically continuous series from olivine basalts to the dark trachytes and rhyolites. The relationship has been confirmed by analysis of their strontium; the ratio of the rare isotope strontium 87 to the common one, strontium 88, in all these rocks is uniform and about the same as that in oceanic basalts. Another index also shows clearly that they were derived from oceanic magmas. The differentiation into the rock varieties in the series apparently came about mainly through gravity separation of components as the original magmas evolved in a series of steps, marked by a distinct iron enrichment in the middle stages. The end products are glassy trachytes and rhyolites, giving evidence that they were produced in a highly fluid state.

We found that a parallel volcanic range in the triangle, the Alayta range, has the same character, structurally and petrographically, as Erta'Ale. To sum up, the petrology of both ranges seems to indicate that they were born as upwellings through fissures in the crust, that their parent material was basaltic magma such as is characteristic of ocean floors, and hence that there is no sialic crust immediately under these ranges. All of this suggests that in the northern part of the Afar triangle continental blocks have already been severed from each other. Evidently this area is a part of the Red Sea ocean-forming system that is separating Arabia from the African continent. Similarly, our survey of the region indicates that the Danakil Alps and two smaller horsts we have detected south of latitude 13 degrees (one close to the Ethiopian escarpment and the other in the middle of the Afar triangle) are continental structures in the process of being split off and separated from the Ethiopian plateau.

Many signs that the northern region of the Afar triangle was covered with seawater in quite recent times turned up in our on-the-ground explorations. On a terrace at the foot of the Ethiopian scarp Faure found a stone axe that was encrusted with seashells, indicating that the sea had covered it after it was abandoned. This axe is from the Acheulean period, not more than 200,000 years ago. We found coral reefs of the Quaternary period (geologically the most recent) in the lava fields of the region. Scattered

about on the floor of the triangle are ash rings, resembling the one at Diamond Head in Hawaii, such as are known to be formed under water; these consist entirely of the shards of volcanic glass called hyaloclastites, which are typical of underwater basaltic eruptions. We even discovered a flat-topped cone, now standing on the dry beds of the triangle, that bears every resemblance to the famous guyots, or flat-topped seamounts, of ocean floors.

The ash rings were particularly interesting to me because I had previously witnessed the formation of two such structures from submarine eruptions, one in the Azores in 1957 and another south of Iceland in 1965. I found that the ash ring is formed as the result of secondary steam explosions from a volcanic eruption under water. The primary eruption tears the molten magma into pieces and hurls them into the water. Because of their large surface-to-volume ratio (a ratio much larger than it would be for a quiet lava flow) the lumps of hot lava transfer enough heat to the water immediately around them to turn it into superheated steam. This steam generates secondary explosions that shatter the lumps into tiny pieces of glass (hyaloclastites) and throw them high into the air—from half a kilometer to more than a kilometer above sea level (which is a great deal higher than material is tossed by the usual volcanic explosions of the Hawaiian or Strombolian type on land). The tiny fragments falling back around the volcano's vent form a large rim that is frequently horizontal and highly regular, because the fragments are deposited under water. We found that the Afar ash rings looked very fresh, and the most eroded one (presumably the oldest) was overlain with corals and shells of marine animals of the Pleistocene period, all of which adds to the evidence that the region was covered by the sea not long ago.

Our examination of the apparent guyot we found in the depression seems to cast a new light on the origin of seamounts. Hundreds of these flat-topped submarine mountains, with their tops in

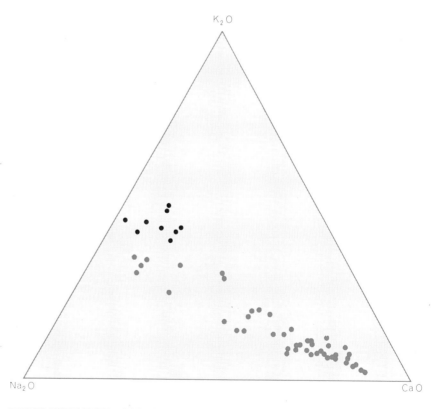

DIFFERENTIATION DIAGRAM shows that the chemical composition of volcanic rock from Mount Erta'Ale (*colored dots*), midway between the Ethiopian escarpment and the Danakil Alps, is different from the composition of the rock from the Pierre Pruvost volcano (*black dots*), which lies close to the escarpment. The Pruvost lavas result either from a "contamination" of deep magmas by the sialic material in the earth's crust or from a melting of the crust itself. The lack of such contamination in the Erta'Ale lavas suggests that, as the Alps and the escarpment have drifted apart, they have split open the earth's crust along a vast rift zone, with the result that no crust remains below the Erta'Ale range.

BLACK RIBBONS of fresh basalt in the Erta'Ale mountains mark zones where molten rock has poured out of fissures in the floor of the Afar triangle. The basalt is chemically similar to the magmas that have welled up from the rifts in the earth's mid-oceanic ridges.

LOW VOLCANIC CONE that marks the northern end of the Erta'-Ale mountain range is further evidence that much of the area was submerged in the recent geologic past. The shards of glass that make up the cone are formed during an underwater eruption as lumps of hot lava turn the water about them into superheated steam. The steam explodes the lumps into hundreds of fragments.

some cases as much as 1,000 fathoms or more below the sea surface, have been discovered in the world ocean. To explain their flat tops, it is generally supposed that the peaks formerly stood above the surface of the water and were eroded flat by the waves, and that the tops are now submerged because the sea bottom sank or the ocean level rose. We found that the truncated cone resembling a seamount in the Afar triangle was built in a way that would account for its flat top simply by the method of its construction. This pile, called Mount Asmara, is about 1,200 feet high and tapers from a width of a mile and a quarter at the base to two-thirds of a mile at the top. It was apparently formed by a buildup of layer on layer of beautiful golden hyaloclastites, which consist of palagonitized olivine basalt and are clearly of submarine origin. Such a process, produced by successive eruptions from a volcanic vent, can account for Mount Asmara's flat top. I am wondering if many of the seamounts in the oceans may not have been built in the same way under water. Perhaps their tops never have in fact been above water but may someday rise out of the ocean as the building process goes on. At all events, it appears that Mount Asmara in the not distant past was totally submerged, with its base more than 1,600 feet below sea level.

If we suppose that the Afar triangle is a part of the Red Sea, the coast of Arabia matches very well the contour of the "coastline" of the part of the African continent from which it is assumed to be separating; the match is at least as good as that between Africa and South America on opposite sides of the Atlantic.

We still have to explain why the axis of the Afar rift is displaced somewhat westward from that of the central ridge in the Red Sea and why the two troughs are now separated. That question will have to await our further explorations in the southern half of the depression. It seems possible that the Gulf of Aden ridge and rift system, thrusting into the depression at right angles to the Red Sea axis, may be exerting a powerful influence that could account for the displacement.

Meanwhile the information obtained so far about the Afar triangle raises an interesting economic question. Because of the absence of sialic crust below the axes of active volcanic ranges of basaltic composition and the probable closeness of the hot mantle to the surface in the northern part of the triangle, a great deal of heat flows into the underground rock strata there. These strata are highly porous and absorb a vast amount of fresh water that drains into the floor of the triangle from the surrounding highlands in the rainy season. Consequently it seems likely that subterranean fields of superheated water and steam underlie parts of this desert region where impermeable strata prevent them from escaping into the atmosphere. If they could be tapped, they might supply millions of kilowatt-hours of cheap electricity per year. This reservoir could supply power to the nearby seaports (Assab, Massawa, Djibouti) to support large new industries (aluminum and other metallurgies, petrochemistry, fertilizers, canneries) in which electricity is the main cost factor. With ores and other raw materials shipped to these ports at low cost, the price of finished products would also be low, and the area could be expected to have a tremendous economic growth. This almost desert region might be an industrial megalopolis in the future—a future far less remote than the geologic one, when the Gulf of Aden and the Red Sea will have expanded into new oceans.

SALT PLAIN lies near the northern apex of the Afar triangle. Each year, when rainfall in the highlands to the east and west drains into this low area, the plain is covered by pools of brine. These are too shallow, however, to hinder the passage of salt caravans.

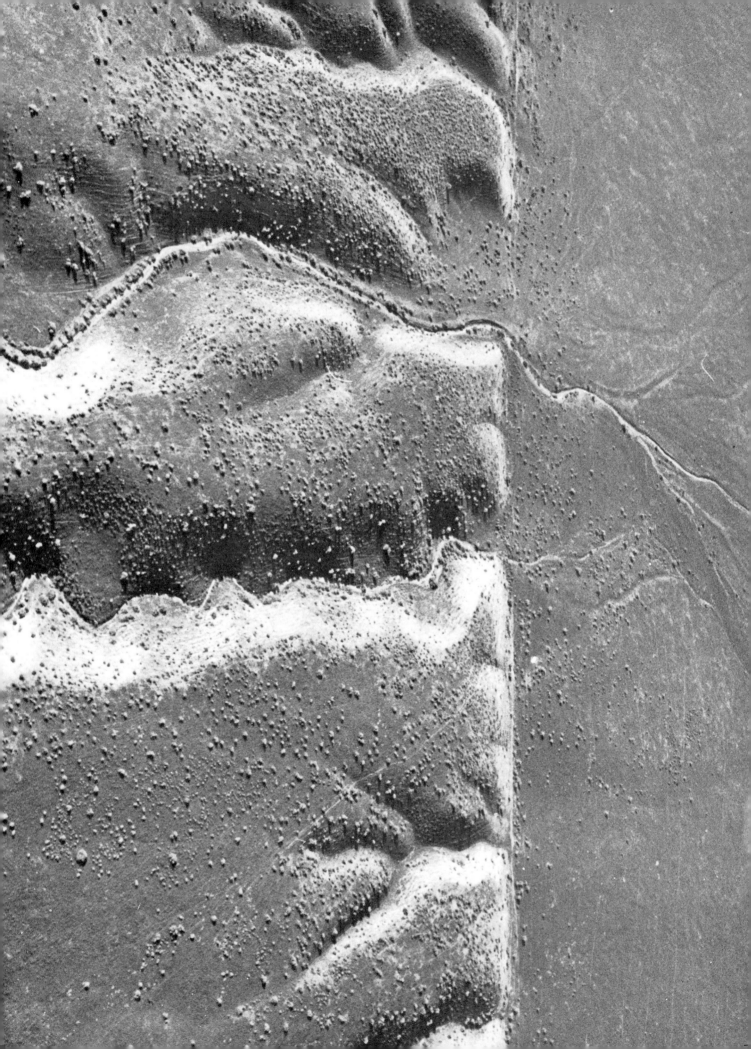

15

THE SAN ANDREAS FAULT

DON L. ANDERSON
November 1971

The San Fernando earthquake that occurred at sunrise on February 9, 1971, jolted many southern Californians into acute awareness that California is earthquake country. Although it was only a moderate earthquake (6.6 on the Richter scale), it was felt in Mexico, Arizona, Nevada and as far north as Yosemite National Park, more than 250 miles from San Fernando. It was recorded at seismic stations around the world. In spite of its relatively small size the San Fernando earthquake was extremely significant because it happened near a major metropolis and because its effects were recorded on a wide variety of seismic instruments. Within hours the affected region was aswarm with geologists mapping faults and seismologists installing portable instruments to monitor aftershocks and the deformation of the ground. It was immediately clear from data telemetered to the Seismological Laboratory of the California Institute of Technology in Pasadena from the Caltech Seismic Network that the earthquake was not centered on the much feared San Andreas fault or, for that matter, on any fault geologists had labeled as active. The faults in the area, however, are all part of the San Andreas fault system that covers much of California.

The San Andreas fault system (and its attendant earthquakes) is part of a global grid of faults, chains of volcanoes and mountains, rifts in the ocean floor and deep oceanic trenches that represent the boundaries between the huge shifting plates that make up the earth's lithosphere. The concept of moving plates is now fundamental to the theory of continental drift, which was long disputed but is now generally accepted in modified form on the basis of voluminous geological, geophysical and geochemical evidence. The theory had received strong support from the discovery that the floors of the oceans have a central rise or ridge, often with a rift along the axis, that can be traced around the globe. Within the rift new crustal material is continuously being injected from the plastic mantle below, forming a rise or ridge on each side of the rift. The newly formed crustal material slides away from the ridge axis. Since the magnetic field of the earth periodically reverses polarity, the newly injected material "freezes" in stripes parallel to the ridge axis, whose north-south polarity likewise alternates. By dating these stripes one can estimate the rate of sea-floor spreading.

The San Andreas fault system forms the boundary between the North American plate and the North Pacific plate and separates the southwestern part of California from the rest of North America. In general the Pacific Ocean and that part of California to the west of the San Andreas fault are moving northwest with respect to the rest of the continent, although the continent inland at least as far as Utah feels the effects of the interactions of these plates.

The relative motion between North America and the North Pacific has been estimated in a variety of ways. Seismic techniques yield values between 1½ and 2½ inches per year. The ages of the magnetic stripes on the ocean floor indicate a rate of about 2⅓ inches per year. Geodetic measurements in California give rates between two and three inches per year. The ages of the magnetic anomalies off the coast of California indicate that the oceanic rise came to intersect the continent at least 30 million years ago. Geologists and geophysicists at a number of institutions (notably the University of Cambridge, Princeton University and the Scripps Institution of Oceanography) have proposed that geologic processes on a continent are profoundly affected when a continental plate is intersected by an oceanic rise. At the rates given above the total displacement along the San Andreas fault amounts to at least 720 miles if motion started when the rise hit the continent and if all the relative motion was taken up on the fault. Displacements this large have not been proposed by geologists, but the critical tests would involve correlation of geology in northern California with geology on the west coast of Baja California, an area that has only recently been studied in detail. One can visualize how the west coast of North America may have looked 32 million years ago by closing up the Gulf of California and moving central and northern California back along the San Andreas fault to fit into the pocket formed by the coastline of the northern half of Baja California. This places all of California west of the San Andreas fault south of

DISPLACEMENT ALONG SAN ANDREAS FAULT is clearly visible in this aerial photograph of a region a few miles north of Frazier Park, Calif., itself 65 miles northwest of Pasadena, where the fault runs almost due east and west. This east-west section of the San Andreas fault is part of the "big bend," where the fault appears to be locked. The photograph is reproduced with north at the right. The hilly region to the left (south) of the fault line is moving upward (westward) with respect to the flat terrain at the right, causing clearly visible offsets in the two largest watercourses as they flow onto the alluvial plain.

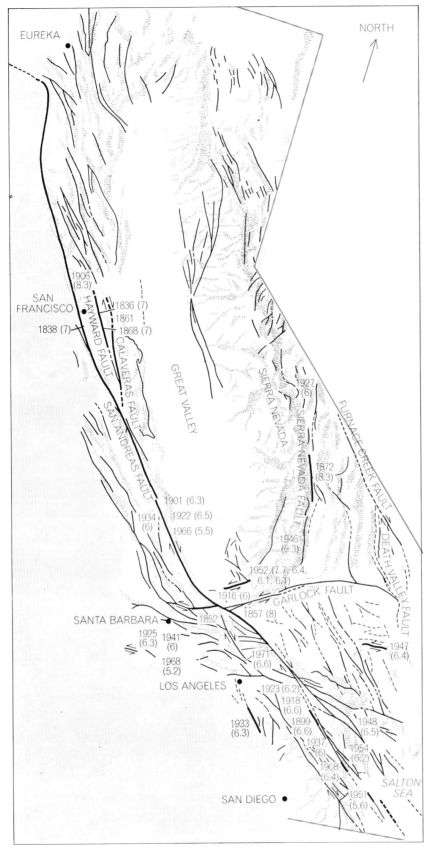

SIMPLIFIED FAULT MAP of California identifies in heavy black lines the faults that have given rise to major earthquakes since 1836. The magnitude of all but two of the earthquakes is given in parentheses next to the year of occurrence. For events that predated the introduction of seismological instruments the magnitudes are estimated from historical accounts. For two major events, the earthquakes of 1852 and 1861, information is too sparse to allow a magnitude estimate. Arrows parallel to the faults show relative motion.

the present Mexican border [*see illustration on page 150*].

California is riddled with faults, most of which trend roughly northwest-southeast, like most of the other tectonic and geologic features of California (such as the Sierra Nevada and the Coast Ranges). The prominent exceptions are the east-west-trending transverse ranges and faults that make up a band some 100 miles wide extending inland from between Los Angeles and Santa Barbara. The San Gabriel Mountains, which form the rugged backdrop to Los Angeles, are part of this complex geologic region, and it was here that the San Fernando earthquake struck. The northeast-trending Garlock fault and the Tehachapi Mountains, which separate the Sierra Nevada and the Mojave Desert, also cut across the general grain of California. The area to the west of most of the northwest-trending faults is moving northwest with respect to the eastern side. This is called right-lateral motion. If one looks across the fault from either side, the other side is moving to the right.

Motion on the Garlock fault is left-lateral, which, combined with the right-lateral motion on the San Andreas fault, means that the Mojave Desert is moving eastward with respect to the rest of California. Parts of the faults that have been observed or inferred to move as a result of earthquakes in historic times are shown in the illustration at the left. Also shown are the dates of the earthquakes and the magnitude of some of the more important ones. In general both the length of rupture and the total displacement are greater for the larger earthquakes. Horizontal displacements as great as 21 feet were observed along the San Andreas fault after the San Francisco earthquake of 1906, which had a magnitude of 8.3 on the Richter scale. (The Richter scale, devised by Charles F. Richter of Cal Tech, is logarithmic. Although each unit denotes a factor of 10 in ground amplitude, or displacement, the actual energy radiated by an earthquake is subject to various modifications.) The San Fernando earthquake produced displacements of six feet, whose direction was almost equally divided between the horizontal and the vertical.

The trend of the San Andreas fault system is roughly northwest-southeast from San Francisco to the south end of the Great Central Valley (the San Joaquin Valley) and again from the north of the Salton Sea depression to the Mexican border. The motion along the faults in these areas is parallel to the fault and is mainly strike-slip, or horizontal. Between these two regions, from the south end of the San Bernardino Mountains to the Garlock fault, the faults bend abruptly and run nearly east and west, producing a region of overthrusting and crustal shortening [*see illustration below*]. The attempt of the southern California plate to "get around the corner" as it moves to the northwest is responsible for the complex geology in the transverse ranges, for the abrupt change in the configuration of the coastline north of Los Angeles and ultimately for the recent San Fernando earthquake. The big bend of the San Andreas fault is commonly regarded by seismologists as being locked and possibly as being the location of the next major earthquake. Much of the motion in this region, however, is being taken up by strike-slip motion along faults parallel to the San Andreas fault and by overthrusting on both

sides of the fault. The displacements associated with the larger earthquakes in southern California in the vicinity of the big bend have averaged out to about 2½ inches per year since 1800. The Kern County earthquake of 1952 (magnitude 7.7) apparently took care of most of the accumulated strain, at least at the north end of the big bend, that had built up since the Fort Tejon earthquake of 1857 (magnitude 8).

The San Andreas fault system cannot be completely understood independently of the tectonics and geology of most of the western part of North America and the northeastern part of the Pacific Ocean. This vast region is itself only a part of the global tectonic pattern, all parts of which seem to be interrelated. The earthquake, tectonic and mountain-building activities of western North America are intimately related to the relative motions of the Pacific and North American plates. Just as it is misleading

to think of the San Andreas fault as an isolated mechanical system, so it is misleading to think of the entire San Andreas fault as a single system. The part of the fault that lies in northern California was activated earlier and has moved farther than the southern California section. The northern portion is less active seismically than the southern section and seems to have been created in a different way. It is also moving in a slightly different direction.

Measuring Displacements

There are several ways to measure displacements on major faults. Fairly recent displacements are reflected in offset stream channels [*see illustration on page 142*]. Many such offsets measured in thousands of feet are apparent across the San Andreas fault in central California, some of which can be directly related to earthquakes of historic times.

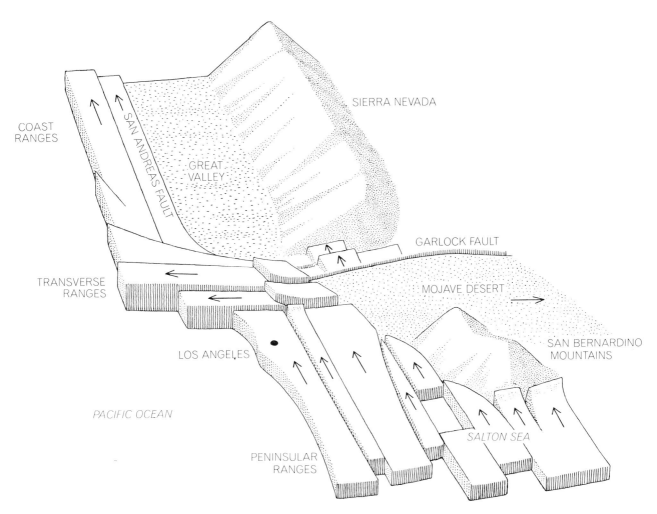

MOTION OF EARTH'S CRUST in southern California is generally northwest except where the lower group of blocks encounters the deep roots of the Sierra Nevada. At this point the blocks are diverted to the left (west), creating the transverse ranges and a big bend in the San Andreas fault system. Above the bend the blocks continue their northwesterly march, carrying the Coast Ranges with them. The Salton Sea trough at the lower right evidently represents a rift that has developed between two blocks.

146

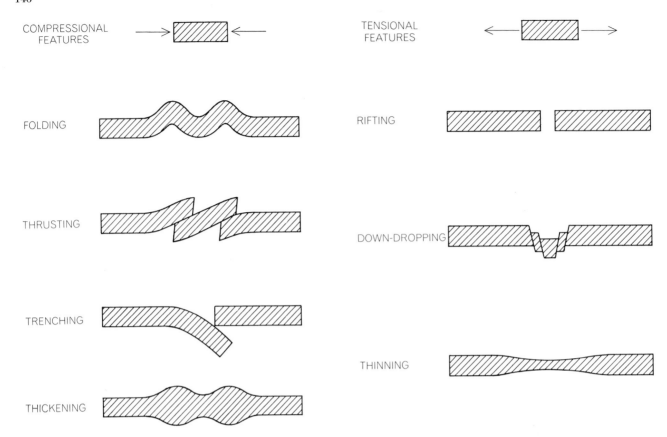

COMPRESSIONAL FEATURES

FOLDING

THRUSTING

TRENCHING

THICKENING

TENSIONAL FEATURES

RIFTING

DOWN-DROPPING

THINNING

RESPONSE OF CRUSTAL PLATES to compression (*left*) and tension (*right*) accounts for most geologic features. According to the recently developed concept of plate tectonics, the earth's mantle is covered by huge, rigid plates that can be colliding, sliding past one another or rifting apart. The rifting usually occurs in the ocean floor. The San Andreas fault marks the location where two plates are sliding past each other. Plate tectonics helps to explain how the continents have drifted into their present locations.

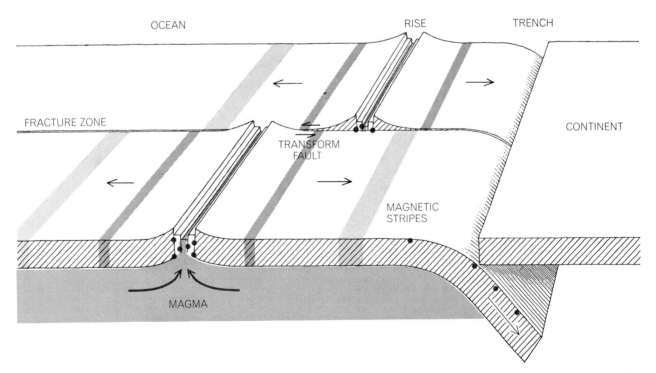

OCEAN

RISE

TRENCH

FRACTURE ZONE

TRANSFORM FAULT

CONTINENT

MAGNETIC STRIPES

MAGMA

RIFT IN OCEAN FLOOR (*color*) initiates three major features of oceanic plate tectonics. The rift is bordered by a rise or ridge created by magma pushed up from the mantle below. The magma solidifies with a magnetic polarity corresponding to that of the earth. When, at long intervals, the earth's polarity reverses, the polarity of newly formed crust reverses too, resulting in a sequence of magnetic "stripes." A trench results when an oceanic plate meets a continental plate. A fracture zone and transform fault result when two plates move past each other. Earthquakes (*dots*) accompany these tectonic processes. The earthquakes in the vicinity of a rise and along a transform fault are shallow. Deep-focus earthquakes occur where a diving oceanic plate forms a trench.

Erosion destroys this kind of evidence very quickly. By matching up distinctive rock units that have been broken up and moved with respect to each other it is possible to document offsets of tens to hundreds of miles. A sedimentary basin often holds debris that could not possibly have been derived from any of the local mountains; matching up these basins with the appropriate source region on the other side of the fault can provide evidence of still larger displacements. When these various kinds of information are combined, one obtains a rate of about half an inch per year for motion on the San Andreas fault in northern and central California over the past several tens of millions of years.

This is much less than the 2½ inches per year that is inferred for the rate of separation of Baja California and mainland Mexico and the rate that is inferred from seismological studies in southern California. There are several possible explanations for the discrepancy. Northern and southern California may be moving at different rates; this seems unlikely since they are both attached to the same Pacific plate. On the other hand, part of the compression in the transverse ranges may result from a differential motion between the two parts of the state. Another possibility is that all of the relative motion between the North American plate and the Pacific plate is not being taken up by the San Andreas fault or even by the San Andreas fault system but extends well inland. The fracture zones of the Pacific seem to affect the geology of the continent for a distance of at least several hundred miles.

The Great Central Valley and the Sierra Nevada lie between two major fracture zones that abut the California coast: the Mendocino fracture zone and the Murray fracture zone. The transverse ranges, the Mojave Desert and the Garlock fault are all in line with the Murray fracture zone. Recent volcanism lines up with the extensions of the Clarion fracture zone and the Mendocino fracture zone. The basins and range geological province of the western U.S., a region of crustal tension and much volcanism, may represent a broad zone of deformation between the Pacific plate and the North American plate proper. Seismic activity is certainly spread over a large, diffuse region of the western U.S.

Although the subject has been quite controversial, most geologists are now willing to accept large horizontal displacements on the faults in California, particularly the San Andreas. Displacements as large as 450 miles of right-lat-

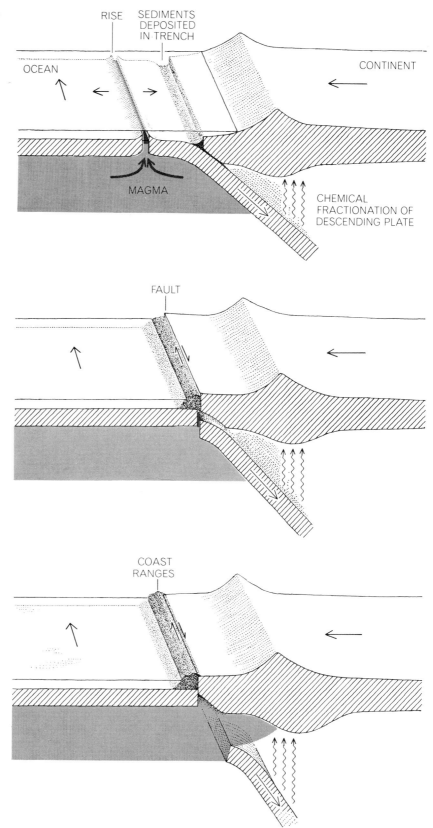

INTERACTION BETWEEN RISE AND TRENCH leads to mutual annihilation. The trench, formed as the oceanic plate dives under the continental plate, slowly fills with sediments carried by rivers and streams (*top*). Meanwhile the melting of the descending slab adds new material to the continent from below. When the axis of the rise reaches the edge of the continent, the flow of magma into the rift is cut off and trench sediments are scraped onto the western (that is, left) part of the oceanic plate (*middle*). The descending plate disappears under the continent and the sediments travel with the oceanic plate (*bottom*). The northern part of the San Andreas fault may have been formed in this way.

eral slip have been proposed for the northern segment of the fault. Displacements on the southern San Andreas fault are put at no more than 300 miles. This discrepancy has been puzzling to geologists. My own conclusion is that the part of northern and central California west of the San Andreas fault has moved northwest more than 700 miles and that the southern San Andreas fault has slipped about 300 miles, which makes the apparent discrepancy even worse. The discrepancy disappears if one drops the concept of a single San Andreas fault and admits the possibility that the two segments of the fault were initiated at different times.

The two-fault hypothesis is supported by straightforward extrapolation of the record on the ocean floor. The two San Andreas faults formed at different times, in different ways and may be moving at different rates. The record indicates that the western part of North America caught up with a section of the East Pacific rise somewhere between 25 million and 30 million years ago. Before the collision a deep oceanic trench existed off the coast such as now exists farther to the south off Central America and South America. The trench had existed for many millions of years, receiving

sediments from the continent; subsequently the sediments were carried down into the mantle by the descending oceanic plate, which was diving under the continent. Based on what we know of trench areas that are active today one can assume that the plate sank to 700 kilometers and that the process was accompanied by earthquakes with shallow, intermediate and deep foci.

The Origin of Continents

Let us examine a little more closely what happens when an oceanic rise, the source of new oceanic crust, approaches a trench, which acts as a sink, or consumer of crust. Evidently the rise and the trench annihilate each other. The oceanic crust and its load of continental debris, which was formerly in the trench, rise because the crust is no longer connected to the plate that was plunging under the continent. The trench deposits are so thick they eventually rise above sea level and become part of the continent. The deposits are still attached to the oceanic plate, however, and travel with it [*see illustration on preceding page*].

In the case of the Pacific plate off California the deposits move northwest

with respect to the continent. This is the stuff of coastal California north of Santa Barbara, particularly the Coast Ranges. According to this view, the northern segment of the San Andreas fault was born at the same time as northern California. The rise and the trench initially interacted near San Francisco, which then was near Ensenada in Baja California. Ensenada in turn was near the northern end of the Gulf of California, which was then closed.

The tectonics and geologic history of California, and in fact much of the western U.S., are now beginning to be understood in terms of the new ideas developed in the theories of sea-floor spreading, continental drift and plate tectonics. Many of the basic concepts were laid down by the late Harry H. Hess of Princeton and Robert S. Dietz of the Environmental Science Services Administration. Tanya Atwater of the University of California at San Diego and Warren Hamilton of the U.S. Geological Survey and their colleagues have made particularly important contributions by applying the concepts of plate tectonics to continental geology. We now know that the outer layer of the earth is immensely mobile. This layer, the lithosphere, is relatively cold and

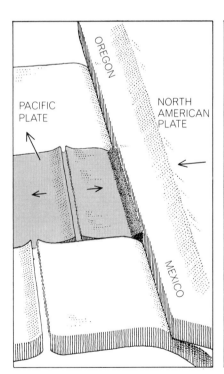
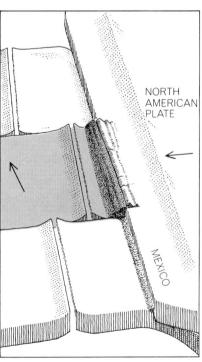
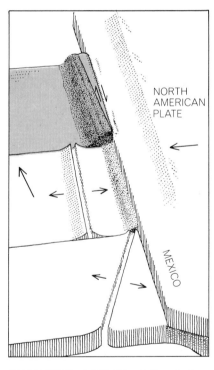

FORMATION OF SAN ANDREAS FAULT SYSTEM is depicted schematically in the six diagrams on these two pages. Some 30 million years ago (*left*) an oceanic rise system lay off the west coast of North America, which was carried by a plate moving toward the rise crests. The continental plate overrides the Pacific plate, producing a long trench. Meanwhile the entire Pacific plate is moving northwest. After a few million years (*right*) the rise nearest the continent is shut off. The trench by now has been filled with material eroded from the continent. These deposits will later become the California Coast Ranges.

NORTHERN SECTION of San Andreas fault is created when the former trench deposits become attached to the northward-moving Pacific plate (*left*). The San Andreas fault lies between the two opposed arrows indicating relative plate motions. Meanwhile to the south a tilted rise crest

rigid and slides around with little resistance on the hot, partially molten asthenosphere.

Where the crust is thick, as it is in continental regions, the temperatures become high enough in the crust itself to cause certain types of crustal rocks to lose their strength and to offer little resistance to sliding. There is thus the possibility that the upper crust can slide over the lower crust and that the moving plate can be much thinner than is commonly assumed in plate-tectonic theory. The molten fraction of the asthenosphere, called magma, rises to the surface at zones of tension such as the mid-oceanic rifts to freeze and form new oceanic crust. The new crust is exposed to the same tensional forces (presumably gravitational) that caused the rift in the first place; therefore it rifts in turn and subsequently slides away from the axis of the rise. In addition to providing the magma for the formation of new crust, the melt in the asthenosphere serves to lubricate the boundary between the lithosphere and the asthenosphere and effectively decouples the two. The rise is one of the types of boundary that exist between lithospheric plates and is the site of small, shallow tensional earthquakes.

When two thin oceanic lithospheric plates collide, one tends to ride over the other, the bottom plate being pushed into the hot asthenosphere. The boundary becomes a trench. When the lower plate starts to melt, it yields a low-density magma that rises to become part of the upper plate; this magma becomes the rock andesite, which builds an island arc on what is to become the landward, or continental, side of the trench. (The rock takes its name from the Andes of South America. Mount Shasta in California is primarily andesite, as are the island arcs behind the trenches that surround the Pacific.) The thickness of the crust is essentially doubled as a result of the underthrusting. The material remaining in the lower plate is now denser than the surrounding material in the asthenosphere, both because it has lost a low-density fraction and because it is colder; thus it sinks farther into the mantle. In some parts of the world the downgoing slab can be tracked by seismic means to 700 kilometers, where it seems to bottom out. By this process new light material is added to the crust and new dense material is added to the lower mantle. A large part of what comes up stays up; a large part of what goes down stays down.

The introduction of chemical fractionation and a mechanism for "unmixing" makes the process different from the one customarily visualized, in which gigantic convection cells carry essentially the same material in a continuous cycle. The new process is able to explain in a convincing way how continents are formed and thickened. As the continent thickens and rises higher because of buoyancy, erosional forces become more effective and dump large volumes of continental sediments into the coastal trenches. A portion of the sediments is ultimately dragged under the continent to melt and form granite. The light granitic magma rises to form huge granitic batholiths such as the Sierra Nevada. A batholith is a large mass of granitic rock formed when magma cools slowly at great depth in the earth's crust. It is carried to the surface by uplifting forces and exposed by erosion.

The concept of rigid plates moving around on the earth's surface and interacting at their boundaries has been remarkably successful in explaining the evolution of oceanic geology and tectonics. The oceanic plates seem to behave rather simply. Tension results in a rise, compression results in a trench and lateral motion results in a transform fault

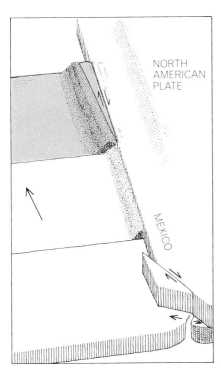

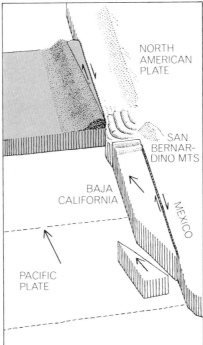

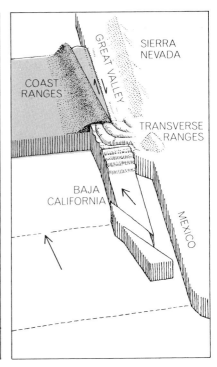

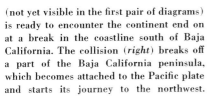

(not yet visible in the first pair of diagrams) is ready to encounter the continent end on at a break in the coastline south of Baja California. The collision (*right*) breaks off a part of the Baja California peninsula, which becomes attached to the Pacific plate and starts its journey to the northwest.

SOUTHERN SECTION of San Andreas fault is now fully activated (*left*) as the Baja California block begins sliding past the North American plate and collides with deeply rooted structures to the north, the Sierra Nevada and San Bernardino Mountains, which deflect the block to the west. More of Baja California breaks loose, opening up the Gulf of California. As Baja California continues to move northwestward (*right*) the Gulf of California steadily widens. The compression at the north end of the Baja California block creates the transverse ranges, which extend inland from the vicinity of present-day Santa Barbara.

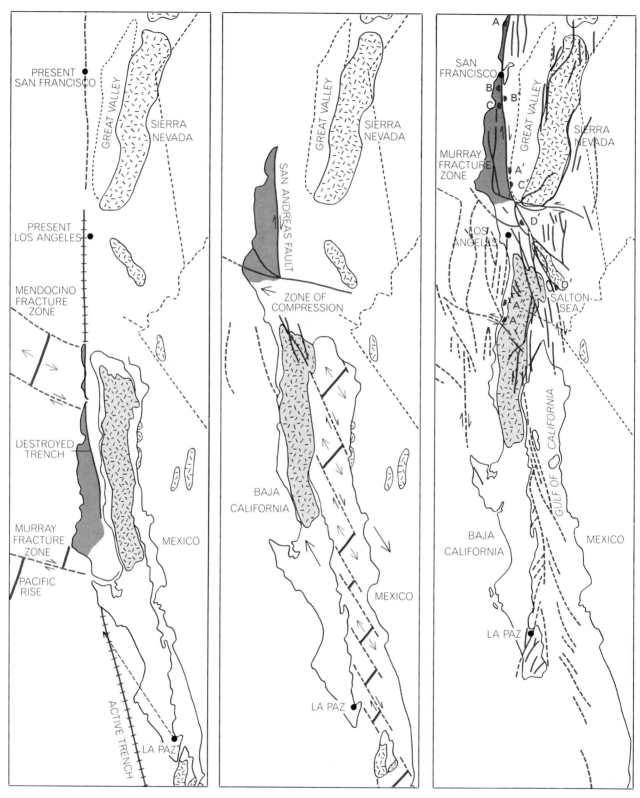

EARLY AND LATE STAGES in the history of the San Andreas fault are depicted. Twenty-five million years ago (*left*) Baja California presumably nestled against mainland Mexico. The first section of oceanic rise between the Murray fracture zone and the Pioneer fracture zone has just collided with the continent. Trench deposits are uplifted and become part of the Coast Ranges of California. The block containing the present San Francisco area (*dark color*) is about to start its long northward journey. A block immediately to the east (*light color*) becomes attached to the Pacific plate and eventually is jammed against the San Bernardino Mountains. Three million years ago (*middle*) the Gulf of California has started to open. As the peninsula moves away from mainland Mexico a series of rifts appear, fill with magma and are offset by numerous fractures. Baja California may have been torn off in one piece or in slivers. Southwest California and Baja California today (*right*) continue to slide northwest against the North American mainland. The illustration shows major fault systems and offshore fracture zones. On the basis of unique rock formations geologists infer that the Los Angeles area has moved northwest about 130 miles (*D'* to *D*) in the past 20 million years or less. Other studies indicate that the Palo Alto region has been carried about 200 miles (*C'* to *C*). Coastal rocks to the north of San Francisco have been displaced at least 300 miles (*A'* to *A*) and perhaps as much as 650 miles (*A"* to *A*) in the past 30 million years.

and a fracture zone [*see bottom illustration on page 146*]. The interaction of oceanic and continental plates or of two continental plates is apparently much more complicated, and this is one reason the new concepts were developed by study of the ocean bottom rather than continental geology.

The boundary between two oceanic plates can be a deep oceanic trench, an oceanic rise or a strike-slip fault depending on whether the plates are approaching, receding from or moving past each other. The forces involved are respectively compressional, tensional and shearing. When a thick continental plate is involved, compression can also result in high upthrust and folded mountain ranges. The Himalayas resulted from the collision of the subcontinent of India with Asia. I shall show that the transverse ranges in California were formed in a similar way. Tension can result in a wide zone of crustal thinning, normal faulting and volcanism; it can also create a fairly narrow rift of the kind found in the Gulf of California and the Red Sea [*see top illustration on page 146*].

The interaction of western North America with the Pacific plate has led to large horizontal motions along the San Andreas fault, to concentrated rifting as in the Salton Sea trough and the Gulf of California, to diffuse rifting and normal faulting accompanied by volcanism in the basins and range province of California, Nevada, Utah and Arizona, to large vertical uplift by overthrusting as in the transverse ranges north and west of Los Angeles, to the generation of large batholiths such as the Sierra Nevada and to the incorporation of deep-sea trench material on the edge of the continent. Ultimately the geology, tectonics and seismicity of California can be related to the collision of North America with the Pacific plate.

Most of the Pacific Ocean is bounded by trenches and island arcs. Trenches border Japan, Alaska, Central America, South America and New Zealand. Island arcs are represented by the Aleutians, the Kuriles, the Marianas, New Guinea, the Tongas and Fiji. The arcs are themselves bordered by trenches. All these areas are characterized by andesitic volcanism and deep-focus earthquakes. Western North America is lacking a trench and has only shallow earthquakes, but the geology indicates that there was once a trench off the West Coast, and in fact there was once a rise. The present absence of a rise and a trench, the absence of deep-focus earthquakes and the existence of uplifted deep-sea sediments are all related.

Tracing back the history of the interaction of the Pacific plate with the North American plate, one is forced to conclude that the northern part and the southern part of the San Andreas fault originated at different times and in different ways. The northern part was evidently formed about 30 million years ago when a portion of rise between the Mendocino fracture zone and the Murray fracture zone approached an offshore trench bordering the southern part of North America. At that time the west coast of North America resembled the present Pacific coast of South America: there was a deep trench offshore, high mountains paralleled the coastline and large underthrusting earthquakes were associated with the downgoing lithosphere.

Origin of the Fault

As the rise approached the continent both the geometry and the dynamics of interaction changed [*see illustrations on pages 148 and 149*]. Depending on the spreading rate of the new crust generated at the rise and the rate at which the rise itself approaches the continent, the relative motion between the rise and the continental plate will decrease, stop or reverse when the rise hits the trench. The forces keeping the trench in existence will therefore decrease, stop or reverse, leading to uplift of the sedimentary material that has been deposited in the trench. In classical geologic terms these are known as eugeosynclinal deposits. Although they have been exposed to only moderate temperatures, they have been subjected to great pressures, both hydrostatic (owing to their depth of burial) and directional (owing to the horizontal compressive forces between the impinging plates). Eugeosynclinal sediments are therefore strongly deformed and become even more so as they are contorted and sheared during uplift. Much of the western edge of California and Baja California is underlain by this material, called the Franciscan formation. The formation is physically attached to the Pacific plate and is therefore moving northwest with respect to the rest of North America. The present boundary is the northern part of the San Andreas fault. Today this section of the San Andreas system extends from Cape Mendocino, north of San Francisco, to somewhere south and east of Santa Barbara, near the beginning of the great bend of the San Andreas fault, where the San Andreas and the Garlock faults intersect.

Meanwhile, 30 million years ago, an-

other section of the rise south of the Murray fracture zone was still offshore, together with an active trench. Baja California was still attached to the mainland of Mexico and the Gulf of California had not yet opened up. The southern part of the San Andreas fault had not yet been formed.

The abrupt change in the direction of the coastline south of the tip of Baja California suggests that here the rise approached the continent more end on than broadside. A sliver of existing continent was welded onto the Pacific plate and rifted away from Mexico, thus forming Baja California and the Gulf of California. Thereafter Baja California participated in the northwesterly motion of the Pacific plate, with the result that the Gulf of California widened progressively with time.

About five million years were required for northern California, which had broken off from Baja California, to be carried about 200 miles to the northwest. At the end of that time the Gulf of California and the Salton Sea trough had not yet opened. The faults that delineate the major geologic blocks in southern California had not yet been activated. The block bearing the San Gabriel fault, now north of San Fernando, occupied the future Salton Sea trough. The transverse ranges will eventually be formed from the Santa Barbara, San Gabriel and San Bernardino blocks by strong compression from the south when Baja California breaks loose from mainland Mexico. This also opens up the Gulf of California and the Salton Sea trough.

As northern California is being carried away from Baja California by the Pacific plate another segment of oceanic rise south of the Murray fracture zone approaches the southern half of Baja California, where the situation described above is repeated except that the rise crest encounters a sharp bend in the coastline and the trench hits just south of the tip of the peninsula. Now instead of approaching the continent more or less broadside the rise approaches the continent end on. Mainland Mexico is still decoupled from that part of the Pacific plate to the west of the rise by the rise and the trench. Baja California, however, is now coupled to the northwestward-moving Pacific plate and Baja California is torn away from the mainland. This happened between four and six million years ago. Magma from the upper mantle wells up into the rift, forming a new rise that works its way north into the widening gulf. Alternatively, the entire peninsula of Baja California could have broken off from the mainland at the

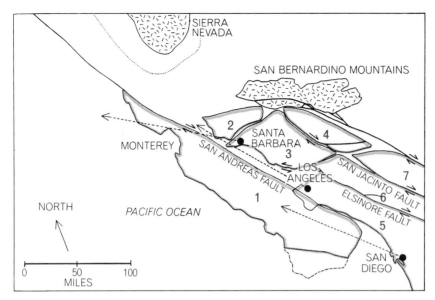

SEQUENCE OF SIMPLIFIED VIEWS shows the movement of major blocks in southern California over the past 12 million years. In the first view (*above*) the Gulf of California has not yet appreciably opened but the block carrying the Coast Ranges (*1*) has started to move rapidly northwest with activation of northern portion of San Andreas fault. Dots show origin and arrows show displacement of San Diego, Los Angeles and Santa Barbara.

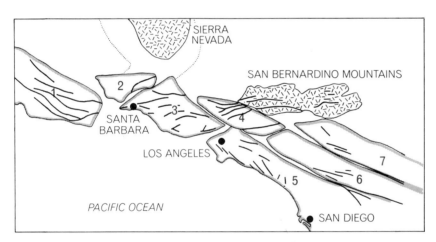

TWO MILLION YEARS AFTER ACTIVATION of the southern portion of the San Andreas fault four blocks (*2, 3, 4, 7*) have been forced against the deep roots of the Sierra Nevada and San Bernardino Mountains. Compressive forces create the transverse ranges. Meanwhile the block carrying the Coast Ranges (*1*) has been carried far to the northwest.

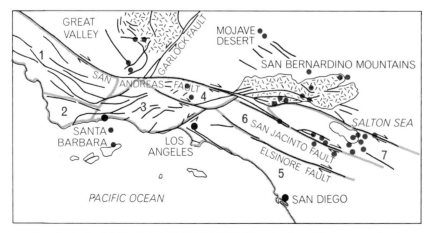

GEOLOGY OF SOUTHERN CALIFORNIA TODAY is dominated by compressive forces operating in the big bend of the San Andreas fault, which connects the southern and northern parts of the system. Colored dots show the location of earthquakes in the recent past.

same time. As the peninsula, including parts of southern California, moves north it collides with parts of the continent that are still attached to the main North American plate. This results in compression, overthrusting and shearing and the eventual formation of the transverse ranges.

The southern part of the San Andreas fault system was therefore formed by the rifting off of a piece of continent. Today it represents the boundary between two parts of the continental plate that are moving with respect to each other. This part of the San Andreas fault was formed well east, or inland, of the southward projection of the northern San Andreas.

The northerly march of southern California and Baja California seems to have been blocked when the moving plate encountered the thick continental crust to the north, particularly the massive granitic San Bernardino mountain range, which includes the 11,485-foot San Gorgonio Mountain. Since large and high mountain ranges have deep roots, the crust in this region is probably much thicker than normal, perhaps as thick as 50 kilometers. Earthquakes in this region are all shallower than 20 kilometers, which may be the thickness of the sliding plates. The blocks veer westward and are strongly overthrust as they attempt to get around the obstacle; this movement generates the big bend in the San Andreas fault system. The deflected blocks eventually join up with the northern California block.

Earthquake Country

From a social and economic point of view earthquakes are one of the most important manifestations of plate interaction. From a scientific point of view they supply a third dimension to the study of faults and the nature of the interactions between crustal blocks, including the stresses involved and the nature of the motions.

Seismologists at the University of California at Berkeley and at the Cal Tech Seismological Laboratory have been keeping track of earthquake activity in California for more than 40 years. Both groups have installed arrays of seismometers that telemeter seismic data to their laboratories for processing and dissemination to the appropriate public agencies. During the 36-year period 1934 through 1969 there were more than 7,300 earthquakes with a Richter magnitude of 4 or greater in southern California and adjacent regions [*see illustration on page 154*]. Many thousands more earthquakes of smaller magnitude are routinely located and reported in the seismological bulletins. Although damage depends on local geological conditions and the nature of the earthquake, a rough rule of thumb is that a nearby earthquake of magnitude 3.5 or greater can cause structural damage. The average annual number of earthquakes of magnitude 3 or greater in southern California recorded since 1934 is 210; the number in any one year has varied from a low of 97 to a high of 391. The strongest earthquake in this period was the Kern County event of magnitude 7.7 in 1952. The aftershocks of that event increased the total number of events for several years thereafter.

In general the larger the earthquake, the greater the displacement across a fault and the greater the length of fault that breaks. The great earthquakes of 1906 and 1857 respectively caused large displacements across the northern and central parts of the San Andreas fault and relieved the accumulated strain in these areas. The accumulation of strain in southern California is relieved mainly by slip on a series of parallel faults and by overthrusting on faults at an angle to the main San Andreas system; that is what happened in the Kern County and San Fernando earthquakes. The unique east-west-trending transverse ranges were formed in this way. In the process deep-seated ancient rocks were uplifted and exposed by erosion.

Another seismically active area associated with major faults is south of the Mojave Desert near San Bernardino, where the faults show a sudden change in direction. The central part of the Mojave Desert is also moderately active. This is consistent with the idea that the sliding lithosphere is diverted by the San Bernardino Mountains. Faults and evidence of relatively recent volcanic events abound in the area. The northern part of Baja California is also quite active. An interesting feature of seismicity maps of southern California is the alignment of earthquakes in zones that trend roughly northeast-southwest, approximately at right angles to the major trend of the San Andreas system.

The map on the next page shows that the San Andreas fault itself has played only a small role in the seismicity of southern California over the past 30-odd years. One must not forget, however, that the great earthquake of 1857 probably broke the San Andreas fault for about 100 miles northwest and southeast from the epicenter. That epicenter is thought to have been near Fort Tejon, which is close to the projected intersection of the Garlock and San Andreas

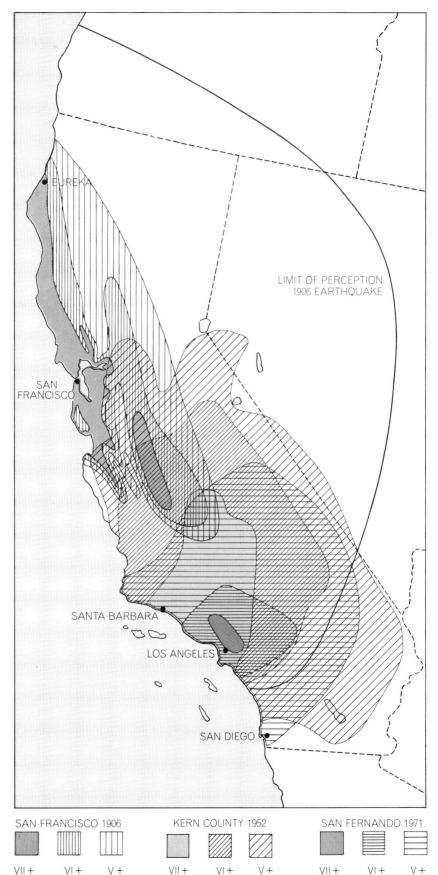

ISOSEISMAL CONTOUR MAP shows the pattern and intensity of ground-shaking produced by the 1906 San Francisco earthquake of magnitude 8.3, the 1952 Kern County earthquake of magnitude 7.7 and the 1971 San Fernando earthquake of magnitude 6.6. The Roman numerals indicate levels of perceived intensities as defined by the modified Mercalli scale. A short description of each level in the scale appears in the text on page 155.

faults; the actual location of the epicenter is uncertain by hundreds of miles because there were no seismic instruments in those days. Since that time this part of southern California has been remarkably quiet and seems to be locked, generating neither earthquakes nor creep. Activity along the San Andreas fault picks up near Coalinga, which is about midway between Bakersfield and San Francisco. Alignments of earthquakes are apparent along the San Jacinto and Imperial faults in the Salton Sea trough near the Mexican border. Although these faults lie west of the main San Andreas fault, they are part of the San Andreas system. The White Wolf fault, which is northwest of and parallel to the Garlock fault, has also been quite active, particularly after

the Kern County earthquake, which occurred on this fault. The White Wolf fault lines up with the Santa Barbara Channel area, which has similarly been quite active.

One way to quantify the seismicity of southern California is to count the number of earthquakes per year per 1,000 square kilometers and compare this figure for the world as a whole. For example, southern California averages one earthquake of magnitude 3 or greater per year per 1,000 square kilometers. Thus within the entire region there are about 200 such earthquakes per year. The rate for earthquakes of magnitude 6.6, the size of the San Fernando earthquake, is about one every five or six years. The actual rates, however, vary

considerably from year to year and depend somewhat on the time interval of the sample. The number of earthquakes decreases rapidly with size, and the average recurrence interval is not well established for the larger earthquakes. Southern California is about 10 times more active seismically than the world as a whole, which is simply to say that California is earthquake country.

Although certain areas in southern California are relatively free of earthquakes, none is immune from their effects. One of the largest quiet areas is the western part of the Mojave Desert wedge. This is surprising because the region is bounded on the northwest and southwest by areas that are obviously under large compression, as is shown by the upthrust mountains in the transverse and Tehachapi ranges and the large overthrust earthquakes that occurred in Kern County and San Fernando. It appears that the region is being protected from the northwesterly march of the southern California–Baja California block by the San Bernardino batholith and may represent a stagnation area in the lee of the mountains. Only a small number of earthquakes are centered near San Diego, although the larger earthquakes in northern Baja California and in the mountains between San Diego and the Salton Sea are felt in San Diego. The Great Central Valley north of Bakersfield and the eastern part of the Sierra Nevada are fairly inactive, as is a large area north of Santa Barbara in the Coast Ranges.

Magnitude and Intensity

It is somewhat deceptive to plot earthquakes as small points on a map. The points represent the epicenter: the point on the surface above the initial break. Once the break is started it can continue, if the earthquake is a major one, for hundreds of miles. Earthquakes of the thrust type, which result from a failure in compression, typically first break many miles below the surface; the surface break and maximum damage can be 10 miles or more from the epicenter. The distance over which strong shocks were felt during three large California earthquakes in this century (1906, 1952 and 1971) can be represented by plotting isoseismals: lines of equal intensity [see illustration on preceding page]. The shape of the pattern varies with the type of earthquake and with the nature of the local geology; structures on deep sedimentary basins or on uncompacted fill get a more intense shaking than structures on bedrock. The isoseismals of the

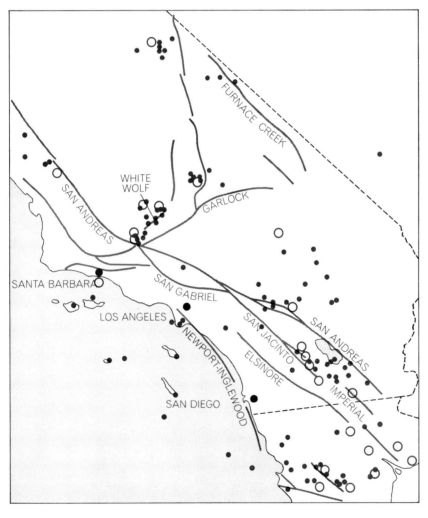

THIRTY-SIX-YEAR EARTHQUAKE RECORD shows the epicenters of all events of magnitude 5 or greater recorded in southern California and in the northern part of Baja California from 1934 through 1969. The epicenter is the point on the earth's surface above the initial break. Dots show earthquakes between 5 and 5.9 in magnitude. Open circles indicate earthquakes of magnitude 6 or greater. The hypocenter, the point of the initial break in the earth's crust, is often many miles below the surface in thrust-type earthquakes, a type frequently observed in this region. In the 36-year period southern California and adjacent regions experienced more than 7,300 earthquakes with a magnitude of 4 or more. Earthquakes are about 10 times more frequent in this area than they are in the world as a whole.

San Francisco earthquake are long and narrow, both because of the orientation of the fault and the length of the faulting and because of the northwest-southeast trend of the valleys. The orientation of the valleys in turn is controlled by the orientation of the San Andreas fault.

The public and the news media are confused about the various measures of the size of an earthquake. There are many parameters associated with an earthquake; they are usually regarded as fault parameters. They include the length, depth and orientation of the fault, the direction of motion, the rupture velocity, the radiated energy, the causal stresses and their orientation, the stress drop (which is related to the strength or the friction along the fault), the energy spectrum, the amount of offset or displacement and the time history of the motion. Most of these parameters can be estimated from seismic records, even from signals recorded several thousand kilometers from the earthquake. To obtain high precision, however, one needs records from many well-distributed seismic stations together with field observations at the site of the earthquake.

The magnitude on the Richter scale is a number assigned to an earthquake from instrumental readings of the amplitude of the seismic waves on a standard seismometer, the Wood-Anderson torsion seismometer. The amplitude must be suitably corrected for spreading and attenuation in the earth, and for instrumental response if a non-standard instrument is used. The magnitude is closely related to the energy of the earthquake, the single most important quantity by which earthquakes can be ranked one against another. If all the corrections are adequately made, a seismologist anywhere in the world will assign the same magnitude. In practice, because of the complicated radiation pattern of earthquakes and because of the distortion of the waves traveling through the earth, the initial magnitude assigned by various observatories may differ slightly. The magnitude scale is logarithmic and is open-ended at both ends. It is not a scale with a maximum value of 10, as is often reported in the press, and negative magnitudes are routinely measured by seismologists working on microearthquakes.

The intensity scale was developed for engineering purposes and is a qualitative measure of the intensity of ground vibration and structural damage. These qualitative assessments are assigned Roman numerals from I to XII. Unlike the magnitude of an earthquake, the inten-

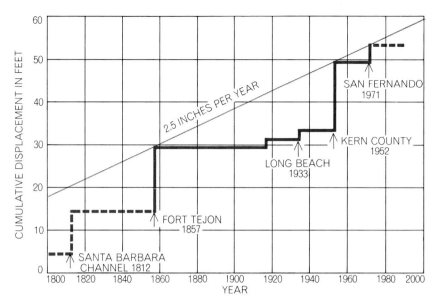

CUMULATIVE DISPLACEMENTS directly related to earthquakes indicate that southern California west of the San Andreas fault system is sliding northwestward at an average rate of 2½ inches per year. Major earthquakes relieve stresses that have built up over decades.

sity varies with distance and depends on the nature of the local ground. In general alluvial valleys, soft sediments and areas of uncompacted fill will magnify ground-shaking and will register higher intensities than adjacent areas on solid rock.

The intensity scale in common usage today is the Modified Mercalli Intensity Scale. The following characterizations of intensity, abridged from longer descriptions, indicate the kind of observations on which the Mercalli scale is based:

I. Not felt except by a very few under special circumstances. Birds and animals are uneasy; trees sway; doors and chandeliers may swing slowly.

II. Felt only by a few persons at rest, particularly on the upper floors of buildings.

III. Felt indoors, but many people do not recognize as an earthquake. Vibrations like the passing of light trucks. Duration of the shaking can be estimated.

IV. Windows, dishes and doors rattle. Walls make creaking sounds. Sensation like the passing of heavy trucks. Felt indoors by many, outdoors by few.

V. Felt by nearly everyone; many awakened. Small unstable objects are displaced or upset; plaster may crack.

VI. Felt by all; many are frightened and run outdoors. Some heavy furniture is moved; books are knocked off shelves and pictures off walls. Small church and school bells ring. Occasional damage to chimneys, otherwise structural damage is slight.

VII. Most people run outdoors. Difficult to stand up. Noticed by drivers of automobiles. Damage is negligible in

buildings of good design and construction, slight to moderate in well-built ordinary structures, considerable in poorly built or badly designed structures. Waves on ponds and pools.

Intensity VII corresponds to the general experience within five or 10 miles of the surface faults associated with the San Fernando earthquake of last February. The following intensity levels were experienced in a small area of the northern San Fernando Valley and would be widely experienced in more severe earthquakes.

VIII. Steering of automobiles affected. Frame houses move on foundations if not bolted down; loose panel walls are thrown out. Some masonry walls fall. Chimneys twist or fall. Damage is slight in specially designed structures, great in poorly constructed buildings. Heavy furniture is overturned.

IX. General panic. Damage is considerable in specially designed structures; partial collapse of substantial buildings. Serious damage to reservoirs and underground pipes. Conspicuous cracks in the ground.

X. Most masonry and frame structures are destroyed with their foundations. Some well-built wooden structures are destroyed. Rails are bent slightly. Large landslides.

XI. Few, if any, masonry structures remain standing. Bridges are destroyed. Broad fissures in the ground. Rails are bent severely.

XII. Damage is nearly total. Objects are thrown into the air.

It is clear that the Mercalli intensity scale is people-oriented; anyone can estimate the intensity from his own experience during an earthquake. The National Oceanic and Atmospheric Administration compiles information on intensities by mailing out questionnaires to a sample of the population living in an area that has experienced a sizable earthquake.

In order to obtain more exact information about the ground motions involved in earthquakes engineers have developed strong-motion accelerometers that automatically trigger and start to record when shaken severely. Most of these instruments are installed in the seismic areas of the U.S., with a particularly heavy concentration in and around Los Angeles. The instruments are expensive and must be located very close to an earthquake to provide useful data. More than 250 of the instruments were triggered during the San Fernando earthquake, and a wealth of engineering data will be provided by these records.

A strong-motion instrument records ground acceleration as a function of time. Accelerations are commonly reported as fractions of a g, the acceleration due to gravity at the earth's surface.

One g is roughly 10 meters per second per second. In designing a building to withstand moderate earthquakes, engineers are concerned chiefly with the maximum accelerations, the period or frequency of shaking and the duration of shaking. Buildings in earthquake-hazard regions with stringent building codes are usually designed to withstand at least .1 g of acceleration; this corresponds to an intensity of about VII on the Mercalli scale.

Although there is no direct correlation between intensity and magnitude, the zone of destruction increases as the magnitude increases for shallow-focus earthquakes. In general the larger the magnitude of an earthquake, the longer the fault length, the larger the displacement across the fault and the longer the duration of shaking. The longer fault length alone accounts for much of the increased area of destruction. For example, the San Francisco earthquake of 1906 had an intensity of VII or greater out to a distance of 500 miles from the epicenter, and this may not have been the largest California earthquake in historic times. The San Francisco earth-

quake had a magnitude of 8.3. The 1952 Kern County earthquake (magnitude 7.7) had an intensity of VII or greater out to 50 miles. The recent San Fernando earthquake (magnitude 6.6) damaged older structures out to 25 miles. An earthquake of magnitude 5.5, the Parkfield earthquake of 1966, produced comparable damage to a distance of 10 miles.

The February Earthquake

The San Fernando earthquake occurred in the San Gabriel Mountains just north of the San Fernando Valley, a densely populated northern suburb of Los Angeles. The San Gabriel Mountains are part of the structural province of the transverse ranges: the band of east-west-trending mountains, valleys and faults that is characterized by strong and geologically recent tectonic deformation. Geologists recognize that the region is one of recent crustal shortening caused by north-south compression. The mountains, produced by buckling and thrusting, are one result of this crustal shortening. They have been thrusting over the valleys to the south for at least five million years along fault planes that dip to the north or northeast.

Although many faults are known to have been active in this area in the past several thousand years, the San Fernando earthquake produced the first historic example of surface faulting. The San Gabriel Mountains rise abruptly some 5,000 feet above the San Fernando Valley and the Los Angeles basin to the south. During the earthquake of February 9 a wedge-shaped prism of the crystalline basement rock comprising the San Gabriel Mountains was thrust over the San Fernando Valley to the southwest, thereby raising the elevation of a section of the San Gabriel Mountains and sliding it slightly to the west. The displacement is consistent with the motions that have been occurring for millions of years, as one can infer from geologic offsets and uplifts. It also agrees with the general picture presented here, namely that the transverse ranges were formed by the collision of the southern and Baja California block with the central and northern California block, and with the concept that the southern California block is being diverted to the west by the massive San Bernardino batholith. One can infer that the thickening of the crust involved in the overthrusting and uplift of the San Gabriel Mountains made this region an additional obstacle to the northwesterly march of

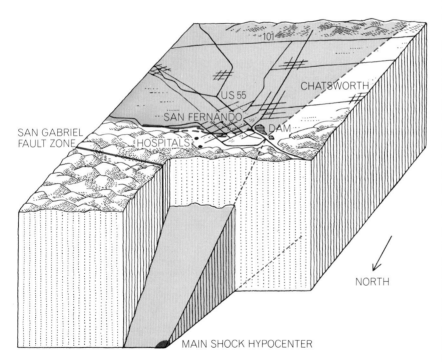

HYPOCENTER OF SAN FERNANDO EARTHQUAKE (*dark color*) of last February was 13 kilometers deep and 12 kilometers north of the area where the principal ground-shaking occurred. The earthquake collapsed sections of two hospitals in the San Fernando Valley, taking 64 lives, and so seriously weakened the earthen wall of the Van Norman Dam at the northern end of the San Fernando Valley that 80,000 people living below the dam had to leave their homes until the water level in the reservoir could be lowered. Total damage caused by the earthquake is estimated at $500 million to $1 billion. This three-dimensional view is based on a drawing prepared by two of the author's colleagues, Bernard Minster and Thomas Jordan, who worked with information supplied by geologists and geophysicists of the California Institute of Technology. The view is looking toward Los Angeles.

the southern California block. If it did, this would lend additional support to the notion that the plates in California are only 15 to 20 kilometers thick. An intriguing possibility is that the upper part of the crust is sliding with relatively little friction on a layer of rock rich in the mineral serpentine.

The hypocenter, or point of initial rupture, of the San Fernando earthquake was at a depth of 13 kilometers under the San Gabriel Mountains. The fault motion was propagated to the surface along a fault inclined northward at an angle of 45 degrees and broke the surface near the cities of San Fernando and Sylmar, at the boundary between the crystalline rocks of the mountains and the sediments of the valleys [*see illustration on page 156*]. Two heavily damaged hospitals were between the epicenter and the surface break and were therefore on the upthrust, or elevated, block. The hundreds of aftershocks following the earthquake covered an area of approximately 300 square kilometers; the total volume of rock lifted up was about 2,500 cubic kilometers.

Even though the elevation difference between the peaks of the San Gabriel Mountains, such as Mount Wilson and Mount Baldy, and the floors of the adjacent valleys is impressive, it does not represent the total uplift. Erosion removes material from the mountains and deposits it in the valleys. The total amount of differential vertical motion probably exceeds two and a half miles, and horizontal displacements in the transverse ranges probably exceed 25 miles. Many thousands of earthquakes of the San Fernando type must have occurred in the area over the past several million years.

Seismic surveillance of the region with instruments dates back only four decades. In this period the northern San Fernando Valley was less active seismically than many other parts of the greater Los Angeles area, although it was comparable to the average for all southern California. On the basis of the seismic data there was no reason to believe the San Fernando area was any more or less likely than any other region of recent mountain-building in southern California to experience a large earthquake. On the other hand, the thrusting and bending associated with the geologic processes in the region, and the tilting that was associated with the earthquake and its aftershocks, suggest that a dense network of tiltmeters could provide a warning of the next large earthquake here.

CONCLUSION

The fifteen articles in this anthology have described the growth of our knowledge of the earth's interior, its ocean floors, and its behavior. Twenty years ago, when Harold C. Urey wrote the first of them, no one knew more than the gross outlines of the structure of the interior nor any details of the structure of the ocean floors. The surface rocks of the continents, however, had been extensively studied for more than a century.

During this short time, the world's oceanographers, led by Maurice Ewing, have mapped the ocean floors and have discovered that their geology is relatively simple and quite unlike that of the continents. Geophysicists, aided by many new seismograph stations, have investigated the interior of the earth and have shown how it enables the surface to break into the moving plates that carry the continents about. The great variety and freedom thus introduced into the study of the earth has resolved many hitherto baffling problems, including the causes of earthquakes and the nature of mountains. Many more problems will be tackled and many details—indeed, most details—still await interpretation. In the future, SCIENTIFIC AMERICAN intends to publish these discoveries as they are made: an article by John Dewey entitled "Plate Tectonics" and one by Sir Edward Bullard entitled "Minerals from the Deep Sea" are being prepared for publication as this goes to press.

One very large question remains unanswered: What is the nature of the forces that move the plates about? Wegener believed that continents raised above the sea floor on a spinning spheroidal earth should be propelled towards the equator. Although this force exists, it is practically negligible, and most authorities, including the authors of several of the preceding articles, have favored convection currents. The nature of these currents has proved to be elusive; if they exist, they must meet many rather strict criteria. Heirtzler has shown, for example, that the South Atlantic has been opening at a rate that has not changed during the past 80 million years, and that the currents, therefore, must be extraordinarily steady and persistent. This is made difficult to understand when we consider that the crust of the midocean ridge must have moved: in Mesozoic time, other continents were clustered around Africa and Antarctica, and so, when breakup began, the midocean-ridge system must have lain around their shores—but it does no longer.

The plates are vast, and they move steadily from midocean ridges towards trenches many thousands of miles away, which suggests a system of convection currents of broad dimensions. However, as Don L. Anderson points out in "The Plastic Layer of the Earth's Mantle," the layer in which such flow must take place is only a few scores of kilometers thick, so the depth of convection cells can be no greater. Theory and experiment both show that convection cells of such shapes are unlikely to occur.

An alternative mechanism that is finding some favor arises from the concept of hot spots—that is, deep-seated upwellings in the mantle—which Wilson and Dietz and Holden mentioned in their articles. Recently, W. J. Morgan proposed that these upwellings of light, hot rock lift the lithosphere over them into domes, each of which is a couple of miles high. Such upwellings may exist under islands along the Mid-Atlantic Ridge, including Jan Mayen, Iceland, Azores, St. Peter and St. Paul Rocks, Ascension, Tristan da Cunha, and Bouvet. If this is so, is it not possible that the crest of this midocean ridge represents a crack that has formed to join the upwellings, and that the Eurasian, African, and American plates are very slowly sliding off the domes in either direction? The same could be true in the other oceans, and all the plates could be sliding together to overlap beneath subduction zones and mountains.

As fantastic as the idea may seem, Anton L. Hales and Wolfgang R. Jacoby have shown that it only requires that the asthenosphere have no permanent strength—gravity will do the rest. If the asthenosphere is indeed weak, such a mechanism could satisfy the features required in the motion of continents—that is, long-lived, steady motion and moving plates of great lateral and small vertical extent.

This mechanism could explain why, in contrast to the uniformity of the South Atlantic and the Pacific, other oceans—for example, the Arctic basins, the Norwegian Sea, Baffin Bay, the Tasman Sea, and the Indian Ocean—have complex patterns and many ridges. If only the southern part of the Mid-Atlantic Ridge and the East Pacific Rise remain steady over their two sets of dominant uplifts, the crests of the other ridges would be forced to migrate off their domes. At intervals, the underlying upwellings might break through to create a new length of active crest and an abandoned old ridge, thus explaining the complexity of these regions.

The third article in Section III, Robert S. Dietz's "Geosynclines, Mountains and Continent-building," was published in SCIENTIFIC AMERICAN just as this book was going to press. (Because it was impossible, at that late date, to include comment upon it in the introduction to that section, it is discussed here.) Dietz relates the new concepts of sea-floor spreading and plate tectonics to the much older realization that there are great prisms of sedimentary rock in the earth's crust. He shows that, after these had accumulated by deltas coalescing and going into continental shelves, subduction zones could have formed beneath them, causing them to be buckled into folded mountains.

Although Dietz suggests that the shelves provide most of the material of which mountains are made (with a moderate addition of magmatic granites), others emphasize the role of volcanoes, which add much andesitic lava derived from subduction zones deep in the mantle. In that connection, it is important to note that, off some coasts, such as that of east Asia, volcanoes form island arcs, whereas along other coasts, such as that of Chile, volcanoes lie on the continent. The difference may be due to different motions of the tectonic plates in relation to the mantle. The articles by J. R. Heirtzler and Sir Edward Bullard have shown that two tectonic plates rotate about a common pole of rotation, but they do not mention that the pole and both plates may be moving together relative to the mantle. Indeed, a consideration of the geometry of the motion of several plates on the globe shows that this situation can be expected to occur. Where it does, the midocean ridge also moves over the mantle; hence, from time to time, the underlying hot spots generate ridges in new locations. I believe that, if the motion is such that a continental crust is advancing over a subduction zone in the mantle, then this may produce a coast, like that of Chile, without island arcs. If the continent is retreating away from a subduction zone, however, island arcs may form offshore and assume the circular form predicted by F. C. Frank.

The uncertainty surrounding these and many other problems shows how much remains to be done. Just as the aftermaths of earlier scientific revolutions have proved to be exciting times in the history of science, so the immediate future promises to be a period of excitement and discovery in the earth sciences.

BIOGRAPHICAL NOTES AND BIBLIOGRAPHIES

[The biographical notes and the bibliographies that originally accompanied these articles in SCIENTIFIC AMERICAN were revised and updated for this anthology.]

I MOBILITY IN THE EARTH

1. The Origin of the Earth

The Author

HAROLD C. UREY won the 1934 Nobel Prize for his isolation of deuterium, the hydrogen isotope with an atomic weight of two; this was a historic development because deuterium provides the simplest instance of the nuclear binding force. Urey is now professor-at-large at the University of California at La Jolla. Born in Walkerton, Indiana in 1893, he received his undergraduate training at the University of Montana and his Ph.D. from the University of California in 1923.

Bibliography

THE PLANETS, THEIR ORIGIN AND DEVELOPMENT. Harold C. Urey. Yale University Press, 1952.

THE ORIGIN OF THE SOLAR SYSTEM. Edited by T. Page and L. W. Page. Mullin, 1966.

THE ORIGIN OF THE SOLAR SYSTEM. H. P. Berlage. Pergamon Press, Inc., 1968.

THE ORIGIN OF THE SOLAR SYSTEM. D. ter Haar in *Annual Reviews of Astronomy and Astrophysics*, Vol. 5, No. 0, page 267; (month) 1967.

ADVENTURES IN EARTH HISTORY. Preston Cloud. W. H. Freeman and Company, 1970.

2. The Trenches of the Pacific

The Authors

At the time their article was written, ROBERT L. FISHER and ROGER REVELLE were members of the University of California's Scripps Institution of Oceanography. Revelle, who had been at Scripps since 1931, was then its director, but he has since moved to Harvard University as Richard Saltonstall Professor of Population Policy and Director of the Center of Population Studies. After graduating from Pomona College in 1929, he went to Scripps as a research assistant, taking his Ph.D. there in 1936. He was a Navy oceanographer during World War II, and was head of the oceanographic section of the task force which conducted the atomic bomb test at Bikini in 1946. Fisher is a marine geologist. He did his undergraduate work at the California Institute of Technology and his graduate work at Northwestern University and at Scripps. "My interest in oceanography," he writes, "dates from a not-particularly-distinguished naval career in the western Pacific in World War II, augmenting early reading of Melville, Maugham and Pierre Loti." For many years, he was engaged in field studies of several Pacific trenches. Since 1960, his primary field interests have been geological and geophysical explorations of the Indian Ocean, to which he has led five major expeditions.

Bibliography

FURTHER NOTES ON THE GREATEST OCEANIC SOUNDING AND THE TOPOGRAPHY OF THE MARIANAS TRENCH. T. F. Gaskell, J. C. Swallow and G. S. Ritchie in *Deep-Sea Research*, Vol. 1, No. 1, pages 60-63; October, 1953.

GRAVITY ANOMALIES AND THE STRUCTURE OF THE WEST INDIES, PART I. Maurice Ewing and J. Lamar Worzel in *Bulletin of the Geological Society of America*, Vol. 65, No. 2, pages 165-173; February, 1954.

THE PHILIPPINE TRENCH AND ITS BOTTOM FAUNA. Anton Fr. Bruun in *Nature*, Vol. 168, No. 4277, pages 692-693; October 20, 1951.

WORLD SEISMICITY MAPS COMPILED FROM ESSA, COAST AND GEODETIC SURVEY, EPICENTRE DATA, 1961-1967. M. Barazangi and J. Dorman in *Bulletin of the Seismological Society of America*, Vol. 59, No. 1, pages 369-380; February, 1969.

THE SEA, VOL. 4, NEW CONCEPTS OF SEA-FLOOR EVOLUTION, PART II, REGIONAL OBSERVATIONS. General Editor, Arthur E. Maxwell. Wiley-Interscience, Inc., 1970.

GEOLOGICAL INVESTIGATIONS OF THE NORTH PACIFIC. Edited by James D. Hays. The Geological Society of America, Memoir 126, 1970.

SEISMIC EVIDENCE FOR THE FAULT ORIGIN OF OCEAN DEEPS. H. Benioff in *Geological Society of America Bulletin*, Vol. 60, No. 12, pages 1837–1856; December, 1949.

EVOLUTION OF THE CENTRAL INDIAN RIDGE, WESTERN INDIAN OCEAN. R. L. Fisher, J. G. Sclater and D. P. McKenzie in *Bulletin of the Geological Society of America*, Vol. 82, No. 3, pages 553–562; March, 1971.

ORIGIN OF PAIRED METAMORPHIC BELTS AND CRUSTAL DILATION IN ISLAND ARC REGIONS. E. R. Oxburgh and D. L. Turcotte in *Journal of Geophysical Research*, Vol. 76, No. 5, pages 1315–1327; February 10, 1971.

STRUCTURAL HISTORY OF THE MARIANA ISLAND ARC SYSTEM. D. E. Karig in *Bulletin of the Geological Society of America*, Vol. 82, No. 2, pages 323–344; February, 1971.

INITIAL REPORTS OF THE DEEP SEA DRILLING PROJECT, NATIONAL SCIENCE FOUNDATION, Vols. 1–5. M. Ewing and others. United States Government Printing Office, Washington, 1970–1971.

3. The Origin of Continents

The Author

MARSHALL KAY is professor of geology at Columbia University. The son of a teacher of geology and mining at the University of Kansas, he took his first degrees at the University of Iowa, then came to Columbia as assistant curator of paleontology. He became assistant professor of geology at Columbia in 1937 and professor in 1944.

Bibliography

GRAVITY ANOMALIES AND ISLAND ARC STRUCTURE WITH PARTICULAR REFERENCE TO THE WEST INDIES. Harry Hammond Hess in *Proceedings of the American Philosophical Society*, Vol. 79, No. 1, pages 71–96; April 21, 1938.

NORTH AMERICAN GEOSYNCLINES. Marshall Kay. Geological Society of America, 1951.

PRECAMBRIAN OF NORTH AMERICA, THE 3RD PENROSE CONFERENCE. P. E. Cloud, Jr. in *Geotimes*, Vol. 16, No. 3, pages 13–18; March, 1971.

PLATE TECTONIC MODELS OF GEOSYNCLINES. W. R. Dickinson in *Earth and Planetary Science Letters*, Vol. 10, No. 2, pages 165–174; 1971.

NORTH ATLANTIC—GEOLOGY AND CONTINENTAL DRIFT, A SYMPOSIUM. Edited by Marshall Kay. The American Association of Petroleum Geologists, Memoir 12, 1969.

PLATE TECTONICS AND GEOSYNCLINES. J. F. Dewey and J. M. Bird in *Tectonophysics*, Vol. 10, Nos. 5 and 6, pages 625–638; 1970.

SEDIMENTARY VOLUMES AND THEIR SIGNIFICANCE. J. Gilluly, J. C. Creed, Jr., and W. M. Cady in *Geological Society of America Bulletin*, Vol. 81, No. 2, pages 353–376; February, 1970.

4. The Interior of the Earth

The Author

K. E. BULLEN, a seismologist, is professor of applied mathematics at the University of Sydney in Australia. Born in Auckland, New Zealand, he was appointed lecturer in mathematics at Auckland University College at the age of 21. It was not until a few years later, he writes, that a conjunction of two events aroused his interest in geophysics: "The first event was the Hawke's Bay earthquake in February, 1931, which was responsible for the greatest loss of life from an earthquake in the history of my home country. The second event was going to England in that same year and chancing to be at the same Cambridge college [St. John's] as Sir Harold Jeffreys, who introduced me to seismological research and inspired me with great enthusiasm for the study. His remarkable stimulation determined my subsequent career." After two and a half years at Cambridge, Bullen returned to his post at Auckland. He then taught at the University of Melbourne for several years, and in 1946 joined the University of Sydney.

Bibliography

THE INTERIOR OF THE EARTH. M. H. P. Bott. Edward Arnold, 1971.

AN INTRODUCTION TO THE THEORY OF SEISMOLOGY, 3rd ed. K. E. Bullen. Cambridge University Press, 1965.

THE EARTH'S MANTLE. Edited by T. F. Gaskell. Academic Press, Inc., 1967.

THE EARTH'S CRUST AND UPPER MANTLE. Edited by P. J. Hart. American Geophysical Union, Geophysical Monograph 13, 1969.

EARTHQUAKES AND EARTH STRUCTURE. J. H. Hodgson. Prentice-Hall, Inc., 1964.

THE EARTH, 5th ed. Sir Harold Jeffreys. Cambridge University Press, 1970.

AN INTRODUCTION TO PLANETARY PHYSICS. W. M. Kaula. John Wiley & Sons, Inc., 1968.

MANTLES OF THE EARTH AND TERRESTRIAL PLANETS. Edited by S. K. Runcorn. Interscience, 1967.

THE APPLICATION OF MODERN PHYSICS TO THE EARTH AND PLANETARY INTERIORS. Edited by S. K. Runcorn. Wiley-Interscience, Inc., 1969.

PHYSICS OF THE EARTH. F. D. Stacey. John Wiley & Sons, Inc., 1969.

5. The Plastic Layer of the Earth's Mantle

The Author

DON L. ANDERSON is professor of geophysics at the California Institute of Technology and director of the Cal Tech Seismological Laboratory. After receiving his bachelor's degree in geology and geophysics at Rensselaer Polytechnic Institute in 1955, he spent a year with an oil company, followed by two years of geophysical research with the Air Force Cambridge Research Center and the Arctic Institute of North America. He received his Ph.D. in geophysics and mathematics from Cal Tech in 1962. His current research activities include seismological investigations of the earth's interior, the physics of the interiors of the moon and terrestrial planets, volcanology, theoretical seismology and studies pertaining to the evolution of the earth and the solar system. In addition he is principal investigator of the Viking seismology experiment that will be landed on Mars in 1976.

Bibliography

THE EARTH AND ITS GRAVITY FIELD. W. A. Heiskanen and F. A. Vening Meinesz. McGraw-Hill Book Company, Inc., 1958.

ELEMENTARY SEISMOLOGY. C. F. Richter. W. H. Freeman and Company, 1958.

LOW-VELOCITY LAYERS IN THE EARTH, OCEAN AND ATMOSPHERE. Beno Gutenberg in *Science*, Vol. 131, No. 3405, pages 959–965; April 1, 1969.

THE ANELASTICITY OF THE EARTH, Don L. Anderson and C. B. Archambeau in *Journal of Geophysical Research*, Vol 69, No. 10, pages 2071–2084; May 15, 1964.

PARTIAL MELTING IN THE UPPER MANTLE, Don L. Anderson and Charles Sammis in *Physics of Earth Planetary Interiors*, Vol. 3, pages 41–50; 1970.

PARTIAL MELTING AND THE LOW-VELOCITY ZONE, Don L. Anderson and Hartmut Spetzler in *Physics of Earth Planetary Interiors*, Vol. 4, No. 1, pages 62–64; December, 1970.

INTRODUCTION TO GEOPHYSICS – MANTLE CORE AND CRUST. George D. Garland. W. B. Saunders Company, 1971.

HOW THICK IS THE LITHOSPHERE? H. Kanamori and Frank Press in *Nature*, Vol. 226, No. 5243, pages 330–311; April 25, 1970.

II CONTINENTAL DRIFT, SEA-FLOOR SPREADING, AND PLATE TECTONICS

6. Continental Drift

The Author
J. TUZO WILSON is professor of geophysics and Principal of Erindale College at the University of Toronto. Wilson first studied at Toronto, taking a B.A. there in 1930, and then at the University of Cambridge, where he received a B.A. and M.A. in 1932. In 1936 he acquired a Ph.D. from Princeton University and for the next three years was an assistant geologist with the Geological Survey of Canada. He served with the Royal Canadian Engineers for the duration of World War II and became professor of geophysics at Toronto in 1946 and Principal of Erindale College in 1967. From 1957 to 1960 Wilson was president of the International Union of Geodesy and Geophysics. A traveler on seven continents, Wilson is the author of *One Chinese Moon*, an account of his month's visit to China in 1958 for the purpose of studying the state of geophysics there, a journey that he repeated in November, 1971.

Bibliography

DID THE ATLANTIC CLOSE AND THEN REOPEN? J. Tuzo Wilson in *Nature*, Vol. 211, No. 5050, pages 676–681; August 13, 1966.

OUR WANDERING CONTINENTS: AN HYPOTHESIS OF CONTINENTAL DRIFTING. Alex. L. Du Toit. Hafner Publishing Company, Inc., 1937.

SUBMARINE FRACTURE ZONES, ASEISMIC RIDGES AND THE INTERNATIONAL COUNCIL OF SCIENTIFIC UNIONS LINE: PROPOSED WESTERN MARGIN OF THE EAST PACIFIC RIDGE. J. Tuzo Wilson in *Nature*, Vol. 207, No. 0, pages 907–911; (month) 1965. Vol. 207, No. 5000, pages 907–911; August 28, 1965.

EVOLUTION OF THE EARTH. R. H. Dott, Jr., and R. L. Batten. McGraw-Hill Book Company, Inc., 1971.

CONVECTION PLUMES IN THE LOWER MANTLE. W. J. Morgan in *Nature*, Vol. 230, No. 5288, pages 42–43; March 5, 1971.

THE STRUCTURE OF SCIENTIFIC REVOLUTIONS., 2nd ed. The University of Chicago Press, 1970.

THE ORIGIN OF THE OCEANIC RIDGES. E. Orowan in *Scientific American*, Vol. 221, No. 5, pages 102–118; November, 1969.

7. The Confirmation of Continental Drift

The Author
PATRICK M. HURLEY is professor of geology at the Massachusetts Institute of Technology. He was born in Hong Kong and lived there until he was nine years old. After being graduated from the University of British Columbia in 1934 with a degree in mining engineering he spent three years mining gold in British Columbia. He obtained a Ph.D. from M.I.T. in 1940 and has been on the faculty there since 1946. He serves as a consultant to industry and government on mineral resources and development. Hurley writes that in addition to his studies of continental drift his recent work "has been on the absolute abundance of minor elements in the earth and on the rate of separation of the crust from the earth's interior."

Bibliography
CONTINENTAL DRIFT. Edited by S. K. Runcorn. Academic Press, Inc., 1962.

THE ORIGIN OF CONTINENTS AND OCEANS. Alfred Wegener, translated by John Biron [from 4th (1929) German edition]. Dover Publications, Inc., 1966.

SPREADING OF THE OCEAN FLOOR: NEW EVIDENCE. F. J. Vine in *Science*, Vol. 154, No. 3755, pages 1405–1415; December 16, 1966.

A SYMPOSIUM ON CONTINENTAL DRIFT. ORGANIZED FOR THE ROYAL SOCIETY. P. M. S. Blackett, F.R.S., Sir Edward Bullard, F.R.S., and S. K. Runcorn in *Philosophical Transactions of the Royal Society of London*, Series A, Vol. 258, pages vii–322; 1965.

CONTINENTAL DRIFT. Ursula Marvin. McGraw-Hill Book Company, Inc., 1971.

UNDERSTANDING THE EARTH. Edited by I. G. Gass, Peter J. Smith and R. C. L. Wilson. Artemis Press, 1971.

BRAZIL-GABON GEOLOGIC LINK SUPPORTS CONTINENTAL DRIFT. G. O. Allard and V. J. Hurst in *Science*, Vol. 163, No. 3865, pages 528–532; February, 1969.

AGAINST THE HYPOTHESIS OF OCEAN-FLOOR SPREADING. V. V. Beloussov in *Tectonophysics*, Vol. 9, No. 6, pages 489–511; June, 1970.

8. Sea-Floor Spreading

The Author

J. R. HEIRTZLER is chairman of the Department of Geology and Geophysics, Woods Hole Oceanographic Institution. He was educated as a physicist, receiving his Ph.D. from New York University in 1953, but has long made hobbies of geology and astronomy. After obtaining his doctorate he taught in universities, worked as a nuclear physicist in industry and, from 1960 to 1967, was in charge of a group that was engaged in research on geomagnetism at the Lamont Geological Observatory of Columbia University. He writes that "firsthand experience with the oceans during World War II and a love of travel combined to give me a strong interest in marine geophysics."

Bibliography

DEBATE ABOUT THE EARTH: APPROACH TO GEOPHYSICS THROUGH ANALYSIS OF CONTINENTAL DRIFT, 2nd ed. H. Takeuchi, S. Uyeda and H. Kanamori, translated by Keiko Kanamori. Freeman, Cooper & Company, 1967.

SPREADING OF THE OCEAN FLOOR: NEW EVIDENCE. F. J. Vine in *Science*, Vol. 154, No. 3755, pages 1405–1415; December 16, 1966.

SEISMOLOGY AND THE NEW GLOBAL TECTONICS, Bryan Isacks, Jack Oliver and Lynn R. Sykes in *Journal of Geophysical Research*, Vol. 73, No. 18, pages 5855–5899; September 15, 1968.

THE DEEP SEA DRILLING IN THE SOUTH ATLANTIC. A. F. Maxwell, R. P. Von Heezen, K. J. Hsu, J. E. Andrews, T. Saito, S. R. Percival, Jr., E. D. Milow and R. E. Boyle in *Science*, Vol. 168, No. 3935, pages 1047–1059; May 29, 1970.

AGE OF THE NORTH ATLANTIC OCEAN FROM MAGNETIC ANOMALIES. W. C. Pitman, III, M. Talwani and J. R. Heirtzler in *Earth & Planetary Science Letters*, Vol. 11, No. 3, pages 195–200; June, 1971.

DISCONTINUITIES IN SEA-FLOOR SPREADING. P. R. Vogt, O. E. Avery, E. D. Schneider, C. N. Anderson and D. R. Bracey in *Tectonophysics*, Vol. 8, Nos. 4–6, pages 285–317; November, 1969.

MAGNETIZED BASEMENT OUTCROPS ON THE SOUTHEAST GREENLAND CONTINENTAL SHELF. P. R. Vogt in *Nature*, Vol. 226, No. 5247, pages 743–744; May 23, 1970.

9. The Origin of the Oceans

The Author

SIR EDWARD BULLARD is professor of geophysics at the University of Cambridge, which he entered as an undergraduate and where he has spent most of his career. He began his professional work there in 1931 as a demonstrator in geodesy. He was an experimental officer with the British Navy during World War II and then returned to Cambridge as a reader in experimental geophysics. In 1948–1949 he was professor of physics at the University of Toronto, and from 1950 to 1955 he was director of the National Physical Laboratory in the United Kingdom. Returning to Cambridge in 1956, he became successively assistant director of research for the university, reader in geophysics and professor of geophysics. He was elected a Fellow of the Royal Society in 1941 and was knighted in 1953.

Bibliography

THE HISTORY OF THE EARTH'S CRUST. Edited by Robert A. Phinney. Princeton University Press, 1968.

SEA-FLOOR SPREADING AND CONTINENTAL DRIFT. Xavier Le Pichon in *Journal of Geophysical Research*, Vol. 73, No. 12, pages 3661–3697; June 15, 1968.

REVERSALS OF THE EARTH'S MAGNETIC FIELD. Sir Edward Bullard in *Philosophical Transactions of the Royal Society of London Series A, Mathematical and Physical Sciences*, Vol. 263, No. 1143, pages 481–524; December 12, 1968.

CONTINENTAL DRIFT: A STUDY OF THE EARTH'S MOVING SURFACE. Don Tarling and Maureen Tarling. Doubleday and Company, Inc., 1971.

HISTORY OF THE EARTH'S MAGNETIC FIELD. D. W. Strangway, McGraw-Hill Book Company, Inc., 1970.

FORMATION OF THE INDIAN OCEAN. W. McElhinny in *Nature*, Vol. 228, No. 5275, pages 977–979; December 5, 1970.

10. The Deep-Ocean Floor

The Author

H. W. MENARD is professor of marine geology at the Institute of Marine Resources and the Scripps Institution of Oceanography of the University of California at San Diego. He received his bachelor's and master's degrees at the California Institute of Technology; in 1949 he obtained his Ph.D. from Harvard University. Since then he has been concerned with marine geology; his work has included participation in 17 oceanographic expeditions, mostly in the Pacific. He has also been a visiting professor at Cal Tech, a Guggenheim Fellow at the University of Cambridge and a staff member of the Office of Science and Technology in Washington. His article is the third he has written for SCIENTIFIC AMERICAN. He is also the author of *Anatomy of an Expedition,* a book describing modern scientific life at sea.

Bibliography
HISTORY OF OCEAN BASINS. H. H. Hess in *Petrologic Studies: A Volume in Honor of A. F. Buddington,* edited by A. E. J. Engel, Harold L. James and B. F. Leonard. The Geological Society of America, 1962.

A NEW CLASS OF FAULTS AND THEIR BEARING ON CONTINENTAL DRIFT. J. Tuzo Wilson in *Nature,* Vol. 207, No. 4995, pages 343–347; July 24, 1965.
SEA FLOOR SPREADING, TOPOGRAPHY, AND THE SECOND LAYER. H. W. Menard in *Transactions American Geophysical Union,* Vol. 48, No. 1, page 217; March, 1967.
RISES, TRENCHES, GREAT FAULTS AND CRUSTAL BLOCKS. W. Jason Morgan in *Journal of Geophysical Research,* Vol. 73, No. 6, pages 1959–1982; March 15, 1968.
EVOLUTION OF TRIPLE JUNCTIONS. D. P. Mackenzie and W. J. Morgan in *Nature,* Vol. 224, No. 5215, pages 125–133; October 11, 1969.
THE FACE OF THE DEEP. B. C. Heezen. Oxford University Press, 1971.

III   SOME CONSEQUENCES AND EXAMPLES OF CONTINENTAL DRIFT

11. The Breakup of Pangaea

The Authors
ROBERT S. DIETZ and JOHN C. HOLDEN are marine geologists with the National Oceanic and Atmospheric Administration, working at the Atlantic Oceanographic and Meteorological Laboratories in Miami. Dietz, who obtained his Ph.D. from the University of Illinois in 1941, has written extensively on geotectonics, geosynclines, continental drift (with the late Harry H. Hess of Princeton he was originator of the sea-floor-spreading concept that is now widely accepted as the underlying mechanism of continental drift), the morphology and structure of the ocean floor, the history of ocean basins, the evolution of continental shelves and slopes, marine mineral resources and astroblemes. Holden is completing work on his doctorate in micropaleontology at the University of California at Berkeley. "As a switch from the megathinking world of plate tectonics and continental drift," he writes, "I find solace in investigating the microcosmic world revealed by the microscope."

Bibliography
CONTINENT AND OCEAN BASIN EVOLUTION BY SPREADING OF THE SEA FLOOR. Robert S. Dietz in *Nature,* Vol. 190, No. 4779, pages 854–857; June 3, 1961.
GEOTECTONIC EVOLUTION AND SUBSIDENCE OF BAHAMA PLATFORM. Robert S. Dietz, John C. Holden, and Walter P. Sproll in *Geological Society of America Bulletin,* Vol. 81, No. 7, pages 1915–1927; July, 1970.
THE FIT OF THE CONTINENTS AROUND THE ATLANTIC. Sir Edward Bullard, J. E. Everett and A. Gilbert Smith in *A Symposium on Continental Drift,* edited by P. M. S. Blackett, Sir Edward Bullard and S. K. Runcorn in *Philosophical Transactions of the Royal Society of London, Series A,* Vol. 258, No. 1088, pages 41–51; October 28, 1965.

PALEOMAGNETISM. E. Irving. John Wiley & Sons, Inc., 1964.
THE FIT OF THE SOUTHERN CONTINENTS. A. Smith and A. Hallam in *Nature,* Vol. 225, No. 5228, pages 139–144; January 10, 1970.
SEA-FLOOR SPREADING. F. J. Vine and H. H. Hess in *The Sea* (Arthur E. Maxwell, General Editor), Vol. 4, Part III, pages 587–622. Wiley-Interscience, Inc., 1970.
THE INDIAN OCEAN AND THE HIMALAYAS—A GEOLOGICAL INTERPRETATION. A. Gansser in *Eclogae Geologicae Helvetiae,* Vol. 59, No. 2, pages 831–848; December, 1966.

12. Continental Drift and Evolution

The Author
BJÖRN KURTÉN is a lecturer in paleontology at the University of Helsinki, where he obtained his Ph.D. in 1954. He has written a number of articles dealing with such subjects as fossil carnivores, dating of early man, late Tertiary and Quaternary stratigraphy, and aspects of population dynamics, evolutionary theory and paleobiogeography. His recent books include *Pleistocene Mammals of Europe* and *The Age of the Dinosaurs.*

Bibliography
VERTEBRATE PALEONTOLOGY. Alfred Sherwood Romer. The University of Chicago Press, 1966.
THE AGE OF THE DINOSAURS. Björn Kurtén. World University Library, 1968.
THE AGE OF MAMMALS. Björn Kurtén. Columbia University Press, 1972.
GONDWANALAND REVISITED: NEW EVIDENCE FOR CONTINENTAL DRIFT. Edited by Gerard Piel in *Proceedings of the American Philosophical Society,* Vol. 112, No. 5, pages 307–353; October, 1968.

TRIASSIC TETRAPODS FROM ANTARCTICA: EVIDENCE FOR CONTINENTAL DRIFT. D. H. Elliot and others in *Science*, Vol. 169, No. 3950, pages 1197–1201; September 18, 1970.

13. Geosynclines, Mountains and Continent-Building

The Author
ROBERT S. DIETZ is a marine geologist with the National Oceanic and Atmospheric Administration, working at the Atlantic Oceanographic and Meteorological Laboratories in Miami. His degrees are from the University of Illinois, where he received his Ph.D. in 1941. He has written extensively on geosynclines, plate tectonics, sea-floor spreading, continental drift, marine mineral resources and deep-research vehicles. "As a departure from investigating the ocean floor," he writes, "I occasionally take trips to various parts of the world to study geologic scars of ancient meteoritic or cometary impacts. These are not circular holes in the ground but usually complex disrupted domes revealing evidence of intense shock." For his research in geotectonics Dietz recently received the Walter H. Bucher medal of the American Geophysical Union and the gold medal of the U.S. Department of Commerce.

Bibliography
NORTH AMERICAN GEOSYNCLINES: THE GEOLOGICAL SOCIETY OF AMERICA, MEMOIR 48. Marshall Kay. Geological Society of America, 1951.
COLLAPSING CONTINENTAL RISES: AN ACTUALISTIC CONCEPT OF GEOSYNCLINES AND MOUNTAIN BUILDING. Robert S. Dietz in *The Journal of Geology*, Vol. 71, No. 3, pages 314–333; May, 1963.
MIOGEOCLINES (MIOGEOSYNCLINES) IN SPACE AND TIME. Robert S. Dietz and John C. Holden in *The Journal of Geology*, Vol. 74, No. 5, Part 1, pages 566–583; September, 1966.
CONTINENTAL MARGINS, GEOSYNCLINES, AND OCEAN FLOOR SPREADING. Andrew H. Mitchell and Harold G. Reading in *The Journal of Geology*, Vol. 77, No. 6, pages 629–646; November, 1969.
MOUNTAIN BELTS AND THE NEW GLOBAL TECTONICS. John F. Dewey and John M. Bird in *Journal of Geophysical Research*, Vol. 75, No. 14, pages 2625–2647; May 10, 1970.
THEORIES OF BUILDING OF CONTINENTS. J. Tuzo Wilson in *The Earth's Mantle*, edited by T. F. Gaskell. Academic Press, 1967.

14. The Afar Triangle

The Author
HAROUN TAZIEFF is with the French Center for National Research. "My childhood was devoted to the idea of becoming a sailor and an arctic explorer," he writes. "Eventually I became a geologist and spent 99 percent of my field time between the Tropics. Once a geologist, I dreamed of high-mountain tectonics and of central Asia; I was confined to old cratons and Africa or Southeast Asia and western America." Tazieff was born in Warsaw and educated in Belgium; he was graduated from the Free University of Brussels in 1932 and has degrees in agronomy and geology from the State Agronomic Institute at Gembloux and the University of Liège respectively. "Besides my interest in investigating the mechanisms of volcanic eruptions and of rift genesis," he writes, "I am fond of Van Gogh's, Gauguin's and some others' painting, of rugby football (by now I am one of the oldest forwards in the world, I think), of knowledge theories, of Stephane Mallarmé's and Robert Vivier's poetry, of mountaineering, of international politics, of the geology of the moon, of architecture and of many other things."

Bibliography
DESERT AND FOREST: THE EXPLORATION OF ABYSSINIAN DANAKIL. L. M. Nesbitt. Jonathan Cape Ltd., 1934.
GEOLOGIA DELL'AFRICA ORIENTALE. Giotti Dainelli. Reale Accademia d'Italia, 1943.
MAJOR VOLCANO-TECTONIC LINEAMENT IN THE ETHIOPIAN RIFT SYSTEM. P. A. Mohr in *Nature*, Vol. 213, No. 5077, pages 664–665; February 18, 1967.
RELATIONS TECTONIQUES ENTRE L'AFAR ET LA MER ROUGE. Haroun Tazieff in *Bulletin de la Société Geologique de France*, Series 7, Vol. 10, No. 4, pages 468–477; 1968.
TRANSCURRENT FAULTING IN THE ETHIOPIAN RIFT SYSTEM. P. A. Mohr in *Nature*, Vol. 218, No. 5145, pages 938–941; June 8, 1968.
TECTONICS OF CENTRAL AFAR. Haroun Tazieff in *Journal of Geology* [in press].
ETHIOPIAN RIFT AND PLATEAUS: SOME VOLCANIC PETROCHEMICAL DIFFERENCES. P. A. Mohr in *Journal of Geophysical Research*, Vol. 76, No. 8, pages 1967–1984; March 10, 1971.
A DISCUSSION ON THE STRUCTURE AND EVOLUTION OF THE RED SEA AND THE NATURE OF THE RED SEA, GULF OF ADEN AND ETHIOPIAN RIFT JUNCTION. N. L. Falcon, I. G. Gass, R. W. Girdler, and A. S. Laughton in *Philosophical Transactions of the Royal Society of London*, Series A, Vol. 267, No. 1181, pages 1–417 (with separate pocket containing 6 charts and maps); October 29, 1970.

15. The San Andreas Fault

The Author
For information about DON L. ANDERSON, see the biographical note for Article 5.

Bibliography
RELATIONSHIP BETWEEN SEISMICITY AND GEOLOGIC STRUCTURE IN THE SOUTHERN CALIFORNIA REGION. C. R. Allen, P. St. Amand, C. F. Richter and

J. M. Nordquist in *Bulletin of the Seismological Society of America*, Vol. 55, No. 4, pages 753–797; August, 1965.

PROCEEDINGS OF CONFERENCE ON GEOLOGIC PROBLEMS OF SAN ANDREAS FAULT SYSTEM. Edited by William R. Dickinson and Arthur Grantz in *Stanford University Publication: Geological Sciences, Vol. XI*. School of Earth Sciences, 1968.

IMPLICATIONS OF PLATE TECTONICS FOR THE CENOZOIC TECTONIC EVOLUTION OF WESTERN NORTH AMERICA. Tanya Atwater in *Geological Society of America Bulletin*, Vol. 81, No. 12, pages 3513–3535; December, 1970.

EARTHQUAKE PREDICTION AND CONTROL. L. C. Pakiser, J. P. Eaton, J. H. Healy and C. B. Raleigh in *Science*, Vol. 166, No. 3912, pages 1467–1474; December, 19, 1969.

EARTHQUAKE PREDICTION AND CONTROL. Allen L. Hammond in *Science*, Vol. 173, No. 3993, page 316; July 23, 1971.

INDEX